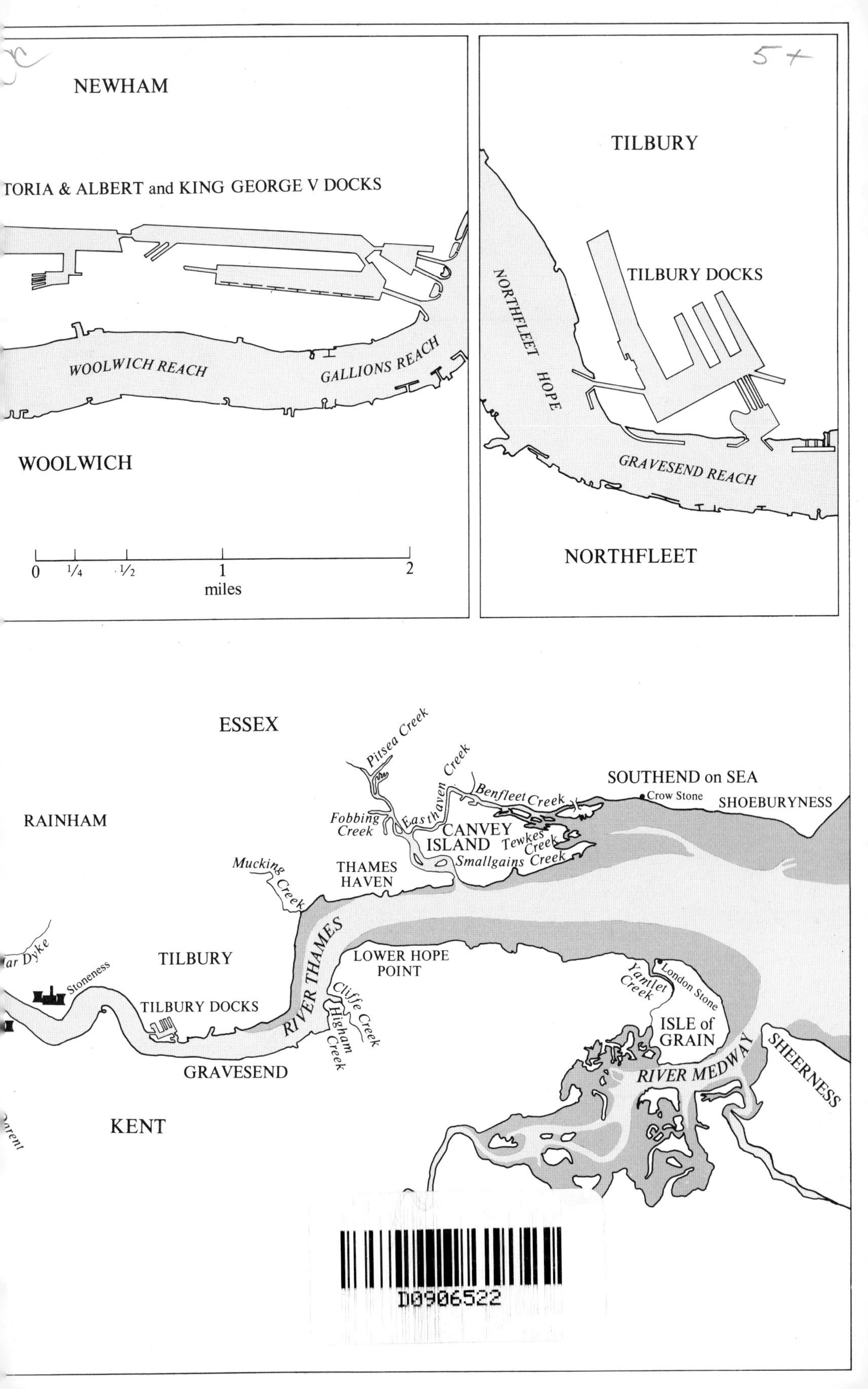
NEWHAM
TORIA & ALBERT and KING GEORGE V DOCKS
WOOLWICH REACH
GALLIONS REACH
WOOLWICH
0
1/4
1/2
1
2
miles
TILBURY
TILBURY DOCKS
NORTHFLEET HOPE
GRAVESEND REACH
NORTHFLEET
ESSEX
Pitsea Creek
East Haven Creek
Benfleet Creek
SOUTHEND on SEA
Crow Stone
SHOEBURYNESS
RAINHAM
Fobbing Creek
CANVEY ISLAND
Tewkes Creek
Smallgains Creek
Mucking Creek
THAMES HAVEN
RIVER THAMES
Stoneness
TILBURY
LOWER HOPE POINT
Yantlet Creek
London Stone
TILBURY DOCKS
Cliffe Creek
Higham Creek
ISLE of GRAIN
GRAVESEND
RIVER MEDWAY
SHEERNESS
KENT

THE TIDAL THAMES

THE TIDAL THAMES

THE HISTORY OF A RIVER AND ITS FISHES

ALWYNE WHEELER

Department of Zoology
The British Museum (Natural History)

ROUTLEDGE & KEGAN PAUL

LONDON, BOSTON AND HENLEY

First published in 1979
by Routledge & Kegan Paul Ltd
39 Store Street, London WC1E 7DD,
Broadway House, Newtown Road,
Henley-on-Thames, Oxon RG9 1EN and
9 Park Street, Boston, Mass. 02108, USA

Set in 11 on 13 Ehrhardt by
Ronset Ltd
Darwen, Lancashire
and printed in Great Britian by
Lowe & Brydone Ltd
Thetford, Norfolk

British Library Cataloguing in Publication Data

Wheeler, Alwyne Cooper

The tidal Thames.
1. Fishes – England – Thames River
I. Title
597'.09204'21 QL633.G7 79-40460

ISBN 0 7100 0200 9

CONTENTS

FIGURES

The illustrations of fishes are taken from William Yarrell's *A History of British Fishes* (third edition, 1859) and were originally engraved by Henry White and Ebenezer Landells, both pupils of Thomas Bewick.

TABLES

INTRODUCTION

The River Thames has been the inspiration of many writers, poets, and artists who have recorded its beauty and its unique history. One of the earliest, Edmund Spenser (1552–99), an Elizabethan poet and a Londoner by birth, penned the immortal lines 'Sweet Thames, run softly, till I end my Song' (*Prothalamion*, 1596), a much quoted line which found later fame in the gifted hands of Robert Gibbings (*Sweet Thames, run softly*, 1940). Since Spenser there have been many books about the Thames, some of which I have been lucky enough to discover and enjoy, and the river seems to be an inexhaustible inspiration to authors.

In presenting yet another book about the river I accept that it will be insignificant beside the literary works of many other authors. However, this book does tell a story that others have not been in a position to tell. First, it is confined to the tidal Thames, from Teddington, where the high weir puts an end to the movement of the North Sea tides, to the mouth of the river, where it gives way to the sea. Second, it tells of a major success in environmental improvement, for much of the river here was until quite recently lifeless as a result of gross pollution, and is now inhabited by a rich fauna of invertebrates and fishes, and attracts large flocks of aquatic birds. The return of fishes and other wildlife to the Thames has captured the attention of naturalists and others all over the world, and has acted as an inspiration to environmental workers in that it shows that a major river can be reclaimed from pollution and that the presence of a great city need not result in harm to the aquatic environment. Citations of this success have appeared in numbers of scientific and popular articles around the world and it is true to say that the revival of life in the Thames has received international attention. Happily too, it has proved an example which has been followed elsewhere, and other rivers, both in the United Kingdom and in other parts of the world, are now showing similar improvements.

It was by good fortune that I was in a position to monitor the changes in the fish fauna of the river from about 1957–8 when there were no fish in the major part of the river until 1973 when it was apparent that a rich fauna had returned and was resident in the Thames. The observations made during my work with the fishes of the tidal river form the major part of this book. The natural history observations are, however, only meaningful when placed in the context of the history of the river and the capital city, London. For this reason, I have hesitantly trespassed into such parts of the social and economic history of London as have a bearing on the water in the river and its tributaries, and more confidently as a naturalist into the past history of the fisheries of the Thames. The history of the factors affecting the fish of the river as London grew and technology changed the way of life of its inhabitants has a fascination of its own and is a story which needs recounting so that the full impact of man on one aspect of the environment can be better appreciated.

My work on the fishes of the Thames would not have been possible without the assistance and co-operation of many people and organisations. Few can be named here, partly on account of the number of people involved but also because some, such as some employees of the various Thames-side electricity generating stations who collected and saved specimens for me, were never known to me by name! Therefore I hope that they will accept a general expression of thanks which is none the less sincere, for without their help it would never have been possible to monitor the return of fishes to the Thames.

Employees of the Port of London Authority, responsible during the period of my work on the Thames for pollution control, were always of the greatest assistance, and amongst them Mr John Potter, Mr Alan Tetlow, and Mr Nigel Baker, all concerned with the control of pollution in the river, never stinted their advice or help. At the Greater London Council, Mr Leslie Wood, distinguished by his deep knowledge of the Thames and London's rivers and their problems and then Assistant Scientific Adviser to the GLC, also gave unstinted help and patiently elucidated the mysteries of sewage treatment to a previously unenlightened naturalist. Capture of fishes in the river was largely effected using some of the

Thames-side electricity generating stations, and it was through the kindness and inspiration of Mr John Wilson and Mr C. M. Gammon of the Station Planning Branch of the Central Electricity Generating Board that these arrangements were made. I owe them my sincere thanks, as I do also to the then superintendents and their staff at the generating stations at Lombard Road and Fulham (now no longer in commission), Blackwall Point, Brunswick Wharf, Littlebrook, and West Thurrock. The General Manager of the Ford Motor Company's works at Dagenham gave permission for his company's power-station staff to collect and save fish, and their help must be acknowledged. Mr Michael Andrews, of the Thames Water Authority, gave me some early information arising from his work with fishes in the river in the period since the Authority was set up; this co-operation is acknowledged.

It is a pleasure to acknowledge the enthusiastic support of the Thames Angling Preservation Society, especially its officers Mr A. E. Hodges and Mr C. Cargill. The TAPS is the oldest conservation body in the country, for unlike many angling societies it owns no water for angling purposes but is simply a body of anglers and naturalists who are dedicated to the preservation of fish life in the Thames, and has been active since 1838, the year of its foundation.

I am also pleased to acknowledge the help in various ways of my wife Cicely and family Tony, Rosalind, and Mary, in particular for their patience while I have been absorbed with writing this book.

I must also acknowledge that most of the laboratory work involved in this study of Thames fishes has been conducted at the British Museum (Natural History) as part of my official duties.

It is necessary to add here that the responsibility for all statements, and any errors of omission or commission are solely my own and in no way reflect the views of any of the above named organisations or individuals.

ONE

THE TIDAL THAMES

The history of the tidal Thames, its fishes, and other wildlife forms an intricate web of inter-relationships between the physical features of an estuary, human endeavour, politics, social history, and biology. As such the full story of the once abundant wildlife of the lower river, its decline, and resuscitation within the past decade can only be appreciated against the wider background. Therefore, although this book is mainly concerned with the fishes and the natural history of the Thames, it is necessary to recount some of the human and other factors which have affected the river's wildlife, for Lord Byron's aphorism 'man marks the earth with ruin' is as true of London's river as it is applicable elsewhere.

PHYSICAL FEATURES

Measuring the extent of the estuary is rather like answering the question 'how long is a piece of string?', except that with an estuary one end can be considered more or less fixed. Today the effect of the tides is observable upstream as far as Teddington Weir, which was until April 1974 (when the Thames Water Authority was created) the boundary between the Thames Conservancy and the Port of London Authority (although the official boundary was, in fact, some distance downstream of the weir). However, it was said by the late A. P. Herbert in *The Thames* (1966), that before the construction of the weir at Teddington the river was tidal as far upstream as Staines, and Staines Stone or the London Stone may mark the upstream limit of the tidal influence. Doubts have been cast on this, and the theory advanced that the stone set up in 1285 may be no more than the marker of the upstream limit of the City of London's jurisdiction. Indeed it has been suggested (Cargill, 1969) that it was no more than the practical limit to which the London Conservators could enforce their charter. However, for all practical purposes the tidal river ends today at the weir at Teddington, 30·4 km (18·9 miles) above London Bridge, although the lock at Richmond at 25 km (15·5 miles) retains the water at a half-tide level during the ebb, a navigational necessity for boats moored between the two locks.

The seaward limit of the estuary is even more difficult to define! Through the accidents of history several administrative limits have been drawn in on maps, although when one sails downstream and out into the North Sea they are scarcely noticeable. The Corporation of London's administrative limit, exerted over navigation, fisheries, and health in medieval times until the Thames Conservancy Act of 1857, was the Crowstone–Yantlet line. The Crowstone stands between Leigh-on-Sea and Southend-on-Sea on the Essex coast, and the London Stone just below Yantlet Creek on the Isle of Grain. By some quirk of administration the City of London is still in charge of the health of the port and maintains an isolation hospital at Denton, near Gravesend. The Thames Conservancy Act of 1857 recognised the same seaward limit until 1894 when the limit was shifted downstream to 50 miles (80·5 km) below London Bridge. Later Acts of 1909 and 1920 brought the Port of London Authority into being, which eventually adopted a seaward limit some 75 miles (120·7 km) below London Bridge.

This limit conveniently coincides with that of the Kent and Essex Sea Fisheries Committee (Harwich to the North Foreland) but this Committee exercises jurisdiction over the fisheries of the whole Essex coast, including the estuary of the Blackwater, and the sea inlets of the River Crouch and Hamford Water behind Walton-on-the-Naze. The creation of the new Water Authorities in 1974 introduced yet a further complication to this as the Anglian Water Authority and the Southern Water Authority now exercise jurisdiction over the tributaries on the Essex and Kent coasts of the estuary. Thus, the Thames estuary has meant and still means different things to different authorities. As neither fishes nor other wildlife have regard for administrative boundaries, perhaps it is not important to set too rigid a seaward limit to observations recorded here. Indeed, one eminent Dutch fishery biologist has even suggested that the whole southern North Sea could be regarded as a type of estuary

(Korringa, 1967) with the Thames, Rhine, Meuse, and Scheldt rivers all forming sub-estuaries. Biologically and, as we shall see, prehistorically this makes sense.

At the end of the last Ice Age the sea level was much lower than it is at present and the greater part of the bed of the southern and central North Sea was exposed as dry land. The Thames and other eastern rivers of England as far north as Yorkshire flowed eastwards and joined with the flow from the Rhine, Meuse, and Scheldt to form a large freshwater lake (Kooijmans, 1972) in the bed of the North Sea. Through this freshwater connection of continental and English rivers many species of freshwater fishes and other native aquatic animals migrated, just as mammals and other terrestrial animals, and plants colonised the land. At a later stage the freshwater overflow from the North Sea lake cut deeper into the chalk barrier in the present Straits of Dover, and as the sea level rose it broke into this channel and the Thames became isolated from the northern English and continental rivers. This is the cause of the slightly less diverse freshwater fish fauna of the Thames when compared with the faunas of the Great Ouse, or river Trent (Wheeler, 1977).

Thus, those Thames freshwater fishes native to the river have a lot in common with those of other southern North Sea rivers, while the marine fauna of the Thames mouth is virtually the same as that of the continental estuaries. Such differences as may exist are due to local physical characteristics and the hand of man.

The Thames estuary (for it is an estuary from Teddington downstream to the North Sea) is like any other estuary in its basic features. It is vaguely bell-shaped, with freshwater entering at the narrow end (over Teddington weir in fact) and at any tributary river mouths that enter along its course. Twice daily the tides flow into the river and its tributaries, except where tidal sluices prevent them entering, the volume of water varying with the amount of freshwater flowing down and with the height of the tide. The tidal stream is very different water to that which flows over Teddington weir; most noticeably it is salt, it is also cooler (at least at the mouth of the river), well oxygenated, heavier than freshwater, and heavily silted. It is this tidal motion with water entering from the seaward end that makes an estuary such a complicated natural system to study, and such a fragile system, for, unlike a freshwater river which flows downstream all the time, in an estuary the same water flows back and forth for days and even weeks, before it reaches the sea. The magnitude of this tidal excursion is seen by the movement of a hypothetical molecule of water at London Bridge at half-tide which was about 7·2 km

(4·5 miles) upstream just after high water and the same distance downstream at low tide (Water Pollution Research Laboratory, 1964). The distances vary between the mouth of the estuary and its source, but an average tidal excursion exists of between 12·9 and 14·5 km (8–9 miles).

As saltwater is denser than freshwater, in the text-book situation where the two meet the freshwater flows over the top of the heavier seawater. Where the two layers come into contact there is mixing with a gradual increase of salinity in the freshwater. The Thames, however, does not conform to the text-book case but is 'well mixed vertically', which is the technical way of saying that on average the salt content declines gradually as one goes further upstream. Probably its nonconformity is caused by the series of looping bends its course follows from Teddington to Gravesend. Because of the effect of friction the velocity of tidal water tends to be greatest near midstream, but at bends it follows the outside curve. The result of the curving course of the river is that the main tidal stream is continually switching from one bank to the other and in the process the fresh and saltwater become thoroughly mixed. This behaviour of the tidal stream has been well known to the Thames watermen for many years, and is the key to the significance of the choice of sides for the University Boat Race each year. Its impact on the pollution of the river and on the distribution of fishes and other aquatic animals was only fully appreciated within the present century.

The salt content of the water does, however, vary with depth on occasions and the surface water is very slightly less salt than the bottom water. In addition, there are areas in the river where the water is not well mixed, especially around St Clement's Reach, where there are two sharp bends in its course at Broadness and Stoneness. Here the tidal stream is forced into an anti-clockwise eddy on the Kent side as the tide flows and on the Essex side as it ebbs. The result is that on the Essex shore the water flows upstream for most of the tidal cycle, and is rather more salt than on the Kent shore. This probably (it has not been proved, but there are indications) affects the distribution of marine fish in the area, and would certainly have an effect on the route of migratory species. Its effect on the overall salinity of the river is not great, but it serves as an example of the complexity of estuarine water flows.

In addition to the freshwater flowing over Teddington Weir, there are a number of tributaries which flow into the tidal Thames. With two exceptions they are not very large. The tributaries are listed briefly here in order of their place of discharge into the river.

The River Crane discharges at Twickenham on the north bank 24·5 km

(15·22 miles) above London Bridge. It rises in the Harrow area of Middlesex and is joined for part of its course by the Duke of Northumberland's River, which is an artificial channel built to take part of the flow from the River Colne. The Duke of Northumberland's River joins the Thames at Isleworth, 24 km (14·9 miles) above London Bridge. Both rivers are small, contributing around 1·7 cubic metres per second (32 million gallons per day) and neither has posed serious pollution problems.

The River Brent rises in the Harrow, Hendon, and Barnet areas and flows into the Thames at Brentford, 22 km (13·7 miles) above London Bridge. For the last part of its course it is joined to the Grand Union Canal, and flows through a canalised channel, ending in navigation locks. Its flow into the Thames is small, 1·1 cubic metres per second (*c.* 21 million gallons per day) (Wood, 1973).

The Beverley Brook enters the main river naturally at Putney Bridge, 12·8 km (8 miles) above London Bridge, although a man-made subsidiary channel flows to near Barnes Bridge three miles upstream. It rises near Cheam in Surrey and flows through urban areas along its length of 29 km (18 miles). Its flow of 0·5 cubic metres per second (9·5 million gallons per day) is composed largely of sewage effluent from two upstream sewage works which, until the late 1960s, were overloaded. By 1971 the quality of this river had improved considerably from its previous rather poor standard (Wood, 1973).

The River Wandle flows into the Thames at Wandsworth, 10·6 km (6·6 miles) above London Bridge and rises near Croydon and Carshalton in Surrey. In its prime it had a considerable reputation as a trout stream and flowed through fashionable rural regions such as Merton. More than most rivers in the London area it suffered at the hand of man and in the 1960s was severely polluted by sewage discharge of low quality, and by industrial effluents, and suffered the final indignity of being classed as a 'metropolitan sewer'. Happily, by the early 1970s it had been restored to some measure of normalcy, and although its flow consists almost entirely of treated sewage effluent, to a volume of 2·1 cubic metres per second (39·9 million gallons per day), it is no longer the serious polluter of the main river that it once was. The improvement in its fauna is a microcosm of the return of animal life to the main river.

Downstream and again on the south bank, the Ravensbourne joins the Thames only 7·2 km (4·5 miles) below London Bridge at Deptford Creek. The Ravensbourne and its tributary, the Quaggy, have a combined length of 70 km (43·5 miles) and flow through densely built-up areas of south London. Its flow into the Thames is small, according to the excellent study by L. B. Wood (1973) being only 0·6 cubic metres per second (11·4 million gallons per day) and of moderate quality in 1971. Like all of the urban rivers of the London area it was moderately severely polluted before the 1970s.

The major natural tributary of the tidal Thames is the Lee (or Lea), which discharges at Bow Creek on the north bank 7·6 km (4·7 miles) below London Bridge. The Lee rises at Luton in Bedfordshire and has several tributaries in Hertfordshire before it joins with the River Stort. In its middle section the Lee is a major source of drinking water extracted to fill the series of reservoirs that stretch between Waltham Abbey and Hackney. In this region its natural course is completely obliterated and the flow of water is through the Lee Navigation Canal (although river-like aqueducts convey water within the reservoir complex). Below the Hackney Marshes the river originally ran through a series of channels; it still does today, but now they are industrialised streams flowing over the original river bed (although the area retains some of the old names in Temple Mills and Three Mills, and the City Mill River, which give clues to their use in earlier days). The lower Lee is linked into the major canal systems of the country through the Hertford Union Canal (to the Grand Union Canal) and the Limehouse Cut (to the Thames at Limehouse between the old Surrey Commercial Docks and the West India Docks). The industrialised and artificial nature of the lower Lee could offer little hope that its water would be of good quality, and most of its flow was composed of treated sewage effluent. Like that of the main river, the water contained little dissolved oxygen, and in the 1950s had high ammonia levels and its temperature was always higher than normal. The aquatic life was poor at this period, but consequent upon improvements at the GLC sewage works at Deephams (near Ponder's End) substantial improvements have taken place in the 1960s and 1970s. The estimated flow at Bow Creek between 1910 and 1960 varied between 42 and 163 million gallons per day (2·2 and 8·6 cubic metres per second), although this is known to be an underestimate (Water Pollution Research Laboratory, 1964).

At Barking Creek the River Roding joins the Thames. It is an Essex river, rising near Great Dunmow and flowing through some of the most beautiful, if modest, country in Essex and indeed giving its name to a series of villages, the Rodings (pronounced Roothings). Its length is some 66 km (41 miles), of which the last 8 km (5 miles) are tidal. Its lower reaches pass through an urban industrialised area and in the 1950s received poor-quality effluents (much of it industrial) from sewage works at Redbridge. In dry weather these and other sewage effluents formed a large part of the Roding's modest flow to the Thames because natural run-off was quick from the mainly arable catchment and a certain amount of abstraction for irrigation purposes took place. At this period the lower Roding was seriously polluted and no fish were found and the invertebrate fauna was impoverished.

Two other Essex rivers enter the Thames within the industrialised region of the Essex shore. The River Beam and the River Ingrebourne join the main river in Halfway Reach and Erith Reach, respectively 22·9 km (14·25 miles) and 24·3 km (15·13 miles) below London Bridge. The Beam rises near Navestock, Essex, and is about 20·1 km (12·5 miles) long. Along its course it receives sewage effluents and industrial waste from the Romford and Hornchurch areas. Near its confluence with the Thames it communicates with Dagenham Breach, so-called because it represents the site of the most serious of many breaches of the sea wall by tide in 1714. Its eventual damming nearly ten years later left a lake of some 55 acres which later became a favoured place for angling, but by the middle of the nineteenth century it was proposed that this be developed as a dock. Its later development as an industrialised region, including the building of the Ford Motor Company's works at Dagenham, was in keeping with the industrialisation of much of the north bank of the Thames and its minor tributaries. The Ingrebourne River is of similar length to the Beam but its source is at Coxtie Green near Brentwood. Its flow is very modest, averaging over a normal year about 6·3 cubic metres per second (120 million gallons per day). In the 1950s it was in poor condition due to the quality of sewage effluent discharged by the works of a local council. About 1 mile of its lower length is tidal and into this the sewage effluent of Dagenham's sewage treatment works is discharged; the overloading of these works at this time meant that Rainham Creek, into which the Ingrebourne flows, was in very poor condition.

At Dartford Creek, 29·3 km (18·19 miles) below London Bridge, the combined flow of the Rivers Darent and Cray enter the Thames. The Darent rises at Limpsfield, Surrey, and the Cray at Orpington in Kent; both rivers flow through pleasant open countryside in their upper reaches and support a rich fauna and flora. They join at Dartford Creek, which represents about 3·2 km (2 miles) of tidal river. Only in this lower part was pollution a problem and oxygen levels were low here in the 1950s, although the combined flow being relatively small the polluting load to the main river was not serious.

The River Mardyke is another minor Essex tributary which enters the

Thames through a sluice at Purfleet 29·5 km (18·35 miles) below London Bridge; its annual average flow is around 0·05 cubic metres per second (1 million gallons per day) but it may vary considerably. Like all the other Thames tributaries it has been used to carry sewage effluent to the main river, and oxygen readings showed a gradual lowering during the period 1930–60.

WATER USE

The uses of the water in the Thames have been many and varied. Carriage of sewage effluent was noted in the previous section to be a primary use of most of the tributary rivers. From the point of view of the condition of the water and the aquatic fauna of the river and its tributaries this is indeed the most important of all the functions of water in the Thames basin, but it is very far from being the only one, and historically it is a relatively recent event.

Water mills represented a basic source of power which has been exploited over a very long period. The Domesday Survey of 1086 recorded that there were about 6,000 water-mills in England, and Wilson (1977) estimates that there were three times this number by the eighteenth century. A water-mill demands that the flowing water is controllable more or less precisely, and the tidal Thames was probably too wide and carried too great a volume of water to be adapted for this use. Upstream, in the non-tidal river, however, numerous water-mills were built, very often at the downstream end of the narrow channel between an island and the river bank, although sometimes a 'cut' or artificial channel would be made to supply water to a riverside mill. Water-mills were numerous in the vicinity of Marlow, Cookham, Taplow, and Hedsor in the Middle Thames valley (Wilson, 1977).

Closer to London, and within the present-day London area, mills were built on many of the tributary streams. Thus, the River Wandle had numerous mills along its length. Its course is not suitable for navigation because of its steepness but this made it ideal for impounding water. By 1086 the Domesday survey showed that there were thirteen corn mills on the Wandle, in 1610 there were twenty-four, four of them being in Wandsworth, and this area was one of the most important flour-milling sites for the London market. The decline in number of corn mills overlapped the period when industries began using water-power, fulling mills were early in the field (1376 at Wandsworth), and dyeing mills (1578 at Wandsworth, 1611 at Carshalton and Wimbledon) which were referred to as Brazil-mills, originally from wood imported from Brazil *via* Portugal (Hobson, 1924). The industrial centre based on the Wandle was early established; foreign immigrants had introduced bleaching and calico-printing there in the seventeenth century, and by 1805 this little river supported twelve calico-printing works, nine flour mills, five snuff mills, three bleaching grounds, and other undertakings such as paper making and leather dressing. The decline of these water-powered industries only came about with the introduction of steam-powered machinery in the nineteenth century (Sheppard, 1971), but some, such as paper making, lingered on until the twentieth century. The Wandle was exceptional (by 1805 it was said to be the hardest-worked river of its size in the world), but other tributary streams carried numerous water-mills. The River Roding at Barking supported at least one water-mill at the time of Domesday (1086), the Barking Abbey 'great mill' which stood at the junction of Barking Creek and the River Roding. Additional mills were built in this vicinity in medieval times, and tanning was another important local industry (Howson, 1975). Similar use was made of the water of the River Lee, both close to its junction with the Thames where its course divided into several smaller channels, some of them artificial cuts, and, of course, upstream through the Lee valley, including the famous gun-powder mills at Waltham Abbey.

Most early water-mills were built at places where it was convenient to control the water's flow to provide power to grind corn. Almost throughout the London area these mills, or new ones built close to them, became power sources for industrial processes, a development which reached a peak in the early years of the industrial revolution. The effects on the tributary rivers were considerable. In most the natural flow was constricted through artificial channels, and where industrial processes were carried out the water was used in addition to providing power. Industrial pollution, although admittedly at first on a small scale, must date from these early tanneries, paper mills, copper mills, and foundries, and other industries which began in the Middle Ages and were to become abundant in the nineteenth century.

The whole of the Thames and most of its tributaries in the tidal reaches have formed a waterway probably since man first settled the riverside terraces at Swanscombe about 250,000 years ago. Like other aspects of water use navigation had little effect on the wildlife of the river until the human populations began to increase dramatically from the Stuart period onwards. Its main interference with the economy of the river was first seen in the tributaries and upstream in the non-tidal region where, due to irregularity of water depth, locks had to be constructed. The earlier flash-locks were often constructed where mill dams penned the water back. When the water was released it provided an exhilarating, if forcible, swoop downstream, but made a wearisome haul for men and horses dragging laden craft upstream against the current and while the flow lasted.

Pound-locks came to the Thames later, although they were in use in Europe in the fourteenth century. They were simply two sets of gates which cut off a small chamber from the river, the water level in which could be raised or lowered so that vessels were spared the often dangerous negotiation of flash-locks. The first pound-lock in the lower Thames basin was built on the River Lee at Waltham Abbey in 1571–4 but it was almost a century later that similar locks were built in the non-tidal Thames at Iffley and Sandford. The importance of river traffic in the eighteenth century must be emphasied. It has been estimated that up to 95 per cent of goods to and from Reading were carried on the river, and Defoe in 1724 in his *Tour thro' . . . Great Britain . . .* makes a special point of detailing the huge quantities of malt, meal, and timber which were carried on barges to the insatiable London market. This trade, especially with London, continued to increase in the nineteenth century; Mr Wilson in his fascinating account *The Making of the Middle Thames* (1977) records that an annual average of 84,000 tons of essential goods were carried on the Thames network, mostly from the Reading area. The Thames Commissioners (first appointed in 1729) were eventually obliged to support this trade with improvements to the navigation, and the building of pound-locks was increased. By 1809 there were twenty-six on the river, and a few years later the Corporation of the City of London built the six locks below Staines and upstream of Teddington. This development of the river for navigation had profound results on the stocks of migratory fishes in the river, and the pound-locks with the high weirs built beside them were eventually to contribute greatly to the extinction of fisheries for salmon, lamperns, and shad (see p. 000).

Below Teddington the Thames was heavily used by shipping, although here they were purely sailing ships, not barges equipped for sailing and towing by gangs of men or horses. Especially below London Bridge they were larger, but the exigencies of navigation in the tidal river, and in particular the form of the old bridge, led to the evolution in the eighteenth century of the majestic and beautiful Thames barges with their fore-and-aft sails and mast stepped in an iron tabernacle on deck so that mast and gear could be lowered as they shot the bridge. Craft like these carried huge quantities of oats, wheat, peas, beans, and fruit from Kent via Rochester on the Medway, as well as raw materials, such as fullers' earth, a rather rare clay-like earth composed mainly of hydrated aluminium silicate, which was used to absorb the grease and to bleach woollen cloth. Much of the fullers' earth came from the famous 'mines' at Boxley and Leeds, near Maidstone, Kent, through the Medway port, a port which also shipped large quantities of wool to London (Marsh, 1971).

The coastal and international trade in and out of London was continuous and heavy, and also led to the fourteenth- and fifteenth-century development of Shadwell, Rotherhithe, and Deptford as shipbuilding and repairing areas. At Deptford the Royal shipyard was set up by Henry VIII in 1513, later to be enlarged and to build many naval ships. The East India Company had a huge shipbuilding yard at Blackwall, while others of their ships were built at Rotherhithe. The Blackwall dock built in 1661 was the first 'wet' dock in London, that is, the ships were built in a basin which could be flooded so as to float them out. In a sense it was the forerunner of all the huge docks that were built between 1805, when the West India Dock was built, to the 1920s, when King George V Dock was built (Docklands Joint Committee, 1975).

The shipbuilding industry on Thames-side sites declined throughout the nineteenth century in the face of competition of yards in the north of England and industries of other kinds developed. Amongst them were the complexes at the Isle of Dogs, Silvertown, and East Greenwich, and the large gasworks at Beckton, built at the end of the century which early on contributed to the serious pollution of the river.

The docks, as such, made little impact on the aquatic fauna, but as they were built on the lower flood plain of the river their impact on the riverside wetlands must have been considerable. At the beginning of the nineteenth century much of the area was still in agricultural use. There were rich pastures on the marshes of the Isle of Dogs, Greenwich, West Ham, and East Ham, and at Blackwall there were complaints by the overseer of the East India Dock that apprentice shipwrights had been known to slip away in pursuit of hares on the marshes. The building of the docks with the associated housing for the men who worked in them, and the development of the riverside by industry all contributed to the overall decline in the quality of the natural aspect of the river, and were a microcosm of the increasingly industrialised region with a startlingly growing human population.

The Thames and its tributaries have always supplied London with its drinking water. While London was still small it virtually supplied itself from wells, rain-water butts, and streams, the former tapping the river gravels and thus providing naturally filtered Thames water of presumably adequate quality. The New River Company (formed in 1609–13) supplied much of the City of London and the western area north of the Thames with good quality water extracted from the River Lee at Amwell, Hertfordshire, and flowing through the 'New River'. The remainder of north-bank London received Thames water pumped up from London Bridge by means of the great water wheel installed in the northernmost arches of the bridge by the Dutchman, Peter Morris, in 1581 (an intake which continued in use until 1821). At the beginning of the eighteenth century new water companies were formed to supplement these services. The Chelsea Water Company formed in 1723, the West Middlesex, the East London, and the Grand Junction sited on the north bank of the river, while the Southwark, the Lambeth, and the Vauxhall companies on the south bank (Sheppard, 1971) supplied Thames water, at first totally untreated, to the surrounding districts. The lower Lee at Old Ford was also a source of supply for the West Ham Company.

Those premises which had a direct water supply were mostly supplied intermittently to basement-level storage tanks from which the water had to be carried to higher floors. In working-class areas the water was supplied, again intermittently, and often at times more convenient to the supplier than the supplied, to common standpipes. The users had then to queue to obtain their water in buckets and other receptacles. With the introduction of steam engines from about 1754 and iron water pipes (as opposed to the hollowed out tree trunks which had been used) the supply was improved and it was possible to deliver water to the upper storeys of buildings. In London as a whole this was a late development, and even the New River Company, one of the best of the companies, and one which supplied the wealthiest areas of London, was pumping at 'high service' to only a quarter of its customers in 1843 (Rose, 1975).

The quality of the water supply was, however, the most vital factor for London and indirectly to the life of the river. With the exception of the New River Company and the supplies from springs on the Bagshot Beds on Hampstead Heath, the water supply companies drew their water from the Thames between Chelsea and London Bridge or the Lee at Old Ford (Sheppard, 1971). By 1852 Parliament made filtration compulsory for all London's water companies, although the Chelsea Company had introduced filter beds as early as 1828. An 1827 parliamentary enquiry was, however, dubious of the efficacy of filtration, commenting that the Thames 'cannot, even when clarified by filtration, be pronounced entirely free from the suspicion of general insalubrity'. Other water companies either failed to filter their water, or the treatment was inadequate, because as late as 1867 the Registrar General summed-up the results of analysis as follows: 'It is much to be regretted that the London waters are not more effectively filtered before distribution. Only on one occasion during the whole year have I obtained a transparent sample of water from the Southwark Company's main' (Rose, 1975). His continuing remarks leave little doubt that some other companies' water was only a little better than this.

A further aspect of water use in the London Thames was of waste disposal. It seems an inbuilt trait of the human species that the quickest way to dispose of some unwanted object is to throw it into the water, a feature which clearly was as true of the Roman inhabitants of London as it is of those today. Admittedly the earlier inhabitants had good cause for this, lacking the vital public services of refuse removal we enjoy today.

In pre-industrial days the Thames was large enough to absorb without damage most of the organic material which was deposited into it, but its smaller tributaries presented a most unsavoury appearance in the city. The Fleet River, a long-extinct tributary, rose in Parliament Hill Fields and flowed to the Thames at present-day Blackfriars Bridge, and was a notorious example of a polluted city river. Its early history included use by a water-mill and it gave passage to small ships from the main river. C. MacKay's *The Thames and its Tributaries* (1840) reported that in 1418 20-ton vessels could go as far upstream as Holborn, some carrying stones and others rushes for floor coverings. The latter were at times a nuisance for they littered the wharves and dropped into the water, contributing to its silting up. But this complaint was already centuries old, for as early as 1290 the White Friars had complained that the stench from the Fleet had caused the death of several of their brethren, and was impossible to deaden with the strongest incense. By 1502 the river had become so choked with silt and rubbish that navigation was impossible and it had to be cleaned out so that ships could continue to take fish and fuel, as well as other goods, as far upstream as Holborn. In addition to the natural siltation of the river as the rain carried dust and rubbish into the water, it represented a convenient disposal place for the waste of slaughter-houses, tanneries, and other industrial refuse. It was also a convenient place to build privies, the seats opening directly over the river from riverside houses, or for public use on the bridges above river. It was to these that Ben Jonson referred in the 1600s, in describing a boat journey up the Fleet, the seat of every privy being 'fill'd with buttock and the walls do sweat urine and plaisters . . .', while each stroke of the oars 'belch'd forth an ayre as hot as the muster of all your night-tubs'. Not surprisingly, the river needed dredging again in 1606 and also in 1652.

The Great Fire in 1666 offered an opportunity to solve the problem of the Fleet, and Hooke and Wren planned and supervised a straighter canalised course. The banks were levelled and new wharves built, tons of silt and the garbage and debris of slaughter-houses and builders' yards were removed and the new canal finished in 1674. Unfortunately, the habits of the population on its bank went unimproved, as Mr Christopher Hibbert in *London. The Biography of a City* (1969) wrote,

> Within a few years it was as choked as the old river had been. The cutlers, dyers, butchers, brewers, tanners and millers, all of whom found it either convenient or necessary to establish themselves beside a stream of water, were making use of it as a rubbish tip.

The *Tatler* in 1710 wrote of,

> Sweepings from butchers' stalls, dung, guts, and blood,
> Drown'd puppies, shaking sprats, all drenched in mud,
> Dead cats, and turnip tops, come tumbling down the flood.

In 1733 the river was arched over with brickwork up to the Fleet Bridge, part was filled in in 1747 (Rudé, 1971), and in 1766 the upper reaches of the river were covered over. The Fleet thus became a sewer, at first running direct to the Thames, but later discharging into one of the main trunk sewers. Its name lingers on in Fleet Street.

On a small scale the decline of this tributary of the Thames foreshadowed the later pollution of the Thames, a history which is recounted in the subsequent chapter.

Up to the end of the eighteenth century the condition of the tidal Thames showed little change from its original state. However, even at that period the stage was set for a major environmental disaster. The ever-increasing prominence of London as a world centre for trade resulted in more shipping, the development of the docks speeded up loading and unloading and thus quicker turn-round of ships; it also called for more manpower, which in turn brought more people to the riverside. The increase in shipping led to greater activity in shipbuilding and related industries, which in turn required more raw material and labour. As the wealth and influence of the city increased so did the numbers of middle-class merchants and managers, creating further demands for goods, clothing, and other materials. The beginning of the

Industrial Revolution, slow and never quite so dramatic in its effect in the south-east of England as in the midlands and the north, also resulted in industry coming to Thames-side with further demands on the river and its environment. All in all, however, the Thames itself did not suffer so severely as its smaller tributaries, such as the Fleet and the Wandle.

While the Great Wen, to use Cobbet's appropriate term, was gathering, it was not to burst until fifty years later. Possibly this respite was one of scale, the human population was still small enough to live without harming the environment too severely. Certainly also it was due to the low level of social demands; drainage was poor, cesspits widespread, the houses and streets not yet lit by gas, and industrial processes still in their early stages of development. One largely fortuitous feature also had a bearing; the many-piered old London Bridge acted in a sense like a tidal barrier. Many of the intakes for drinking water were upsteam of the Bridge, while much of the industry and the working areas were downstream. Upstream of the Bridge freshwater was pounded back, downstream it was strongly tidal with all the resulting special problems of polluted water in the ebb and flow of a tidal estuary. Thus the City and Westminster could enjoy a cleaner and certainly better-looking river than the inhabitants of, for example, Deptford. Although the public health aspects of city life were not remarked on at this period (and waited for the pioneering Edwin Chadwick in the 1830s) there was probably a marked difference in the incidence of waterborne disease between west and east London to which old London Bridge significantly contributed.

TWO

POLLUTION OF THE RIVER AND REMEDIES

The condition of London's river has for long been a matter of concern to those who have responsibility for it, or who simply had to live and work beside it. As early as 1357 Edward III complained 'that dung and other filth had accumulated in divers places upon the banks of the river and . . . fumes and other abominable stenches arising therefrom' and initiated an attempt to clean the area (Howard, 1975). This was one of the earliest recorded complaints about the condition of the river, a subject that was ever more frequently raised by writers in the eighteenth and nineteenth centuries. Tobias Smollett (1721–71), well known as a ferocious novelist, but also a widely experienced and humane, Scottish-trained surgeon, recognised the appalling quality of Thames water in his immortal *Humphrey Clinker* (1771):

> If I would drink water, I must . . . swallow that which comes from the river Thames, impregnated with all the filth of London and Westminster. Human excrement is the least offensive part of the concrete, which is composed of all the drugs, minerals, and poisons, used in mechanics and manufacture, enriched with the putrefying carcases of beasts and men; and mixed with the scourings of all the wash-tubs, kennels, and common sewers, within the bills of mortality.

While Smollett may have exaggerated its condition to some extent there is little doubt that the river was becoming seriously polluted. As we have seen, up to the end of the eighteenth century the worst pollution was concentrated in the smaller streams running through the densely populated areas. Thus, the Fleet had already a long history of pollution and noisome conditions, and no doubt, the other city rivers such as the Walbrook, and to the west the Tyburn, had comparable conditions. In a sense the problem with London's rivers was one of scale, and the ever-increasing population began to make more and severer demands on its environment.

THE HUMAN ELEMENT

In Smollett's day, despite his references to drugs, minerals, and poisons used in manufacture, the principal pollutant was organic matter, mainly human excrement, but also the waste from the 150 city slaughter-houses, the fish market, tanneries, and domestic rubbish. Whether it was dumped from riverside premises or disposed of into the drains and streams which ran to the river, the end result was that it or its liquid effluent reached the river and began to decompose. In the process of decomposition bacteria remove oxygen from the water (or the air if on land) until the organic matter is reduced to basic compounds, such as carbon dioxide, ammonia, phosphates, and others, of which the first two are more or less poisonous to animal life-processes. So, organic pollution is in effect a double-headed monster in that it takes oxygen from the water and produces toxic substances. Again the question of scale comes into effect. A small quantity of rotting material is quickly absorbed, the oxygen being replaced by both that gained where the water is in contact with air, especially where the surface is disturbed as at weirs, riffles, or waterfalls, and by oxygen produced by aquatic plants and algae during photosynthesis. The toxic substances are dispersed through a large volume of water and some even serve to nourish a healthy growth of plants and algae. A river is thus self-purifying to a great extent. However, when the volume of organic matter increases beyond the capacity of the river to absorb, neutralise, and purify, then pollution is the problem. So much oxygen may be used by the micro-organisms that the water becomes anaerobic (deprived of oxygen). A deposit of highly organic mud begins to form on the bed whenever the current allows and this is both unpleasant to see and worse to smell, because another group of bacteria obtain their oxygen first from the available nitrates and then reduce the contained sulphates to produce hydrogen sulphide, which has a pungent 'bad-egg smell' and turns the water black due to the precipitation of insoluble iron sulphide. This kind of situation is symptomatic of serious organic pollution and was reached in the London Thames during the middle of the nineteenth century. Its advent was undoubtedly hastened because the river here is tidal (see p. 6).

A major contributory factor was the growth of London. In 1700, the combined population of the City, the Borough (Southwark), Westminster, the out-parishes of Middlesex and Surrey, and the parishes of Marylebone, St Pancras, Hammersmith, Kensington and Chelsea, approximated to 674,350 (Rudé, 1971). Social pressures kept the first three relatively stable in total, but by 1801 the out-parishes and the five above-named parishes had expanded remarkably, so that the comparable total was 900,000. By 1811 the number had risen to about 1,100,000, of whom only one-tenth lived within the City, and by 1855 it had swollen to well over 2 million (Sheppard, 1971). Some of this growth was accounted for by the westward spread from Westminster, especially the development of Belgravia by Thomas Cubitt, beginning in 1825 by raising the level of the lowlying, marshy ground with soil excavated from St Katherine's Dock to the east of the Tower of London. Development of the south bank of the river had been slow and mainly confined to the Borough, in the vicinity of London Bridge, in which area the City exercised certain rights, but between 1816 and 1819, three new bridges were built, Vauxhall, Waterloo, and Southwark, plus new turnpike roads which resulted in a sudden increase in building and population there. This increase was to continue almost all round the nucleus formed by the City and Westminster through the nineteenth century, but at an ever-increasing rate. The population of the London area swelled from over 2 million in 1850 to 4,750,000 in 1880.

This dramatic population increase led indirectly to the gross pollution of the tidal Thames as well as its inner London tributaries, one reason being the introduction of that boon of civilised life, the water closet. By long-standing practice, if not tradition, domestic sewage had been held in cesspits which were periodically cleaned out, the accumulated night-soil being removed by nightmen for eventual disposal to market gardens and farms around the city. This was the ideal situation but one which fell short in practice, especially in the poorer parts of London where cesspits went unemptied for years on end. Much refuse also found its way into the streets, which were not notably well drained, and this, plus the overflows from flooded cesspits, eventually drained into the ditches and street sewers which communicated with the river. While the human population remained relatively small and stable such practices were not entirely harmful, although in many instances they were insanitary as, for example, where a well was sunk in close proximity to the cesspit. However, with the increase in population, especially in areas where large numbers of people lived in poor conditions, problems abounded and serious threats were posed to public health. This situation was much aggravated with the introduction of water closets, which became popular with the middle classes from about 1830, although they had first been installed in the London area some twenty years earlier. In 1843 Thomas Cubitt, the builder, claimed that 'there are 10 water-closets put up now where there was only one 20 years ago' (Rose, 1975). By 1850 the areas served by the most efficient water companies, the Chelsea and New River companies, had on average one closet for every two houses, although elsewhere the average was one to every five or six houses (Rose, 1975). In the working-class areas up till the 1840s communal cesspits were the general rule, and water closets were unknown. It is also probable that some water closets fed into cesspits rather than the sewers, with consequent flooding and dispersal of pathogenic organisms. However, sewers existed, although many functioned more as storage for waste-matter rather than conduits, and in 1848 the newly formed Metropolitan Commission of Sewers, motivated by public health considerations, began a vigorous programme, to abolish cesspools, clear the sewers, and improve house drainage. Edwin Chadwick (1800–90), the great social reformer and one of the Commissioners, was convinced that regular flushing of the sewers with water was the most efficacious way of cleansing (Sheppard, 1971), and this was adopted as the policy of the Commission, with disastrous results to the condition of the Thames.

NINETEENTH-CENTURY INDUSTRIAL POLLUTION

Industrial pollution contributed further to the deterioration of the river, but in general was not as serious a problem as the organic pollution caused by waste of a large human population. However, in the eyes of several authors it was in part responsible for deterioration of the river and its fisheries. William Yarrell, author of *A History of British Fishes* (1836) claimed that 'the numerous gas and other manufactories on the banks of the river . . . have affected the quality of the water', and Venables (1874) was even more damning of gas-works and the decline of salmon in the river. Venables wrote:

> I suggest, however, that the emptying of all sorts of injurious matter into the Thames is the chief and increasing source of evil,

> and that amongst these none are so bad as the gas-works; indeed, a few facts connected with gas-works tally with remarkable accuracy with the records of salmon-fishing at Boulter's Lock. Thus gas-works began to be used in London about 1813, and the very next year the salmon are diminishing in size as well as number!

The direct association that Venables suggested between the introduction of gas-works and the decline of the salmon is obviously an over-simplification (the varied causes of the decline are discussed later p. 58), but there seems to be little doubt that gas-works did make a considerable contribution to the pollution of the river.

Street gas-lighting came to London in July 1807 when the proprietors of the Golden Lane Brewery illuminated both Golden Lane and Beech Street, and later that year when the publicity-minded Frederick Albert Winsor, a German national, illuminated the south side of fashionable Pall Mall with thirteen lamps. Winsor formed a National Light and Heat Company following this first successful venture and from this grew the Gas Light and Coke Company by Act of Parliament in 1810. The company at once began to contract to light the main thoroughfares in Westminster and other public places, making the coal-gas at a plant in Great Peter Street in Westminster. Other plants were opened to serve small areas and it is clear that gas-making was approached in a piecemeal fashion, small plants being built to serve each area. This was doubly apparent when between 1817 and 1825 six more supplying companies were established in London and another four companies between 1830 and 1836 (Chandler and Lacey, 1949). The Metropolis Gas Act of 1860 reduced the competition between the rival companies and demarcated the districts each was to serve, and in the same year the original chartered Gas Light and Coke Company under new directors assumed a new lease of life which was later to lead it to pre-eminence in London gas supply. The chartered company began building in 1868 a new and much enlarged works beside Gallions Reach, near Barking, which was to emerge as Beckton Gas Works, in its time the largest gas-works in Europe and the centre of London's gas supply until the 1970s when natural gas became available.

The very number of small works in the early days of gas supply in London indirectly contributed to pollution of the Thames. At first, their prime function was simply to produce coal-gas, the resultant waste, although it did have some value, being disposed of as expeditiously as possible. Most, if not all, the works were sited close to the river or its navigable tributaries for the simple reason that the raw material, coal, was carried by boat. Thus the river was the natural recipient of much of the waste matter, whether by intent or by accidental leakage. The production of gas from coal was effected by strongly heating it in the absence of air, and producing in the process coal-tar, ammoniacal liquor, and coke. Much of the coal-tar and ammoniacal liquor (gas-liquor) was removed by passing the gas through water, either through a water-filled seal above the retort or through water-cooled condensers, thus producing a liquid residue of extreme toxicity, but it was not until the 1830s that a flourishing market developed for the by-products, of which tar for the navy was the most important (Sheppard, 1971). However, around 1817 research had shown that a more effective method of removing the impurities from coal-gas was by passing it through 'cream of lime', a creamy mixture of lime in water, and this method was used for many years (Chandler and Lacey, 1949). The lime, when taken out of the purifier, was dumped in heaps and the smell of the 'sulphur compounds' it contained was so objectionable that late in the eighteenth century the 'spent lime nuisance' was subject of a special report to the Local Government Board. No doubt much of the spent lime found its way to the river, thus contributing to the polluted state of the water.

Even when an outlet had developed for some of the by-products, the residual waste retained compounds of acute toxicity. Jones (1964) lists the following components of ammoniacal gas-liquor, 'free ammonia, ammonium salts, cyanide, sulphide, thiocyanate and a variety of aromatic compounds including pyridine, phenols, cresols and xylenols'. Even when treated to remove the ammonia the resulting 'spent gas-liquor' contains phenol, or carbolic acid, which is 'probably the most dangerous to fish' (Jones, 1964).

It can be seen therefore that the numerous small gas-making plants which were scattered along the Thames in the first half of the nineteenth century were a source of pollutant to the river of a quite exceptional toxicity. There can be little doubt that they contributed materially to the extinction of the more sensitive of the species of fish and other animals, and they can hardly have improved the quality of the river as a source of drinking water! This indeed was the conclusion of a correspondent to the *Field* (17 August 1861) who wrote,

> The beautiful river Wandle has been for some years at times polluted to so great an extent that annually a large number of fine trout are destroyed. Within the last week upwards of a hundred of

> these fish have been taken out dead at Mitcham, and the water is so affected by gas-tar, or some such matter, that it is totally unfit to drink.

As the gas industry developed technologically and the numerous small companies amalgamated to build larger gas-works many of the toxic by-products were removed and marketed, but even as late as 1882 the giant gas-works at Beckton was discharging waste of a highly toxic nature (see p. 38).

CHOLERA AND THE DISPOSAL OF SEWAGE

London's sewers, like its early gas-woıks, had grown in a piecemeal fashion, each parish being self-sufficient and the City entirely independent. Until the nineteenth century the word sewer signified a channel for the removal of surface water, and it was not until the growth of London's population, the introduction of water closets, and the impetus of public health reformers that sewers began to assume their present-day connotation and function. Indeed, until about 1811 the connection of bog-houses, or houses of office, to sewers was formally forbidden in London, although only occasionally actually prevented (Sheppard, 1971). Metropolitan drainage was administered by eight independent Commissions of Sewers, some of whom acted under the powers of the original Act of Parliament of 1532, and all of whose sewers discharged to the Thames. In Westminster until 1815 it was an offence to connect private drains to the Commissioners' sewers, while in Tower Hamlets householders were permitted to make such connections but they were charged for doing so.

The great movement towards public health legislation, led by Edwin Chadwick, a pupil and disciple of Jeremy Bentham, eventually produced some kind of organisation to this chaotic situation which posed a threat to, the health of the inhabitants of London. The largely humanitarian motives which inspired Chadwick were reinforced by a powerful ally choelra. Although the causes of cholera were not then understood, it had become obvious to Chadwick and others, that the regions where this and other diseases struck most severely were where sanitation was poorest and overcrowding most intense. As *The Times* remarked, 'The cholera is the best of all sanitary reformers, it overlooks no mistake and pardons no oversight'. It was through this stimulus that the Metropolitan Commission of Sewers was established in 1848, with extensive powers throughout London, although not in the City, which had comparable legislation in the

City Sewers Act of 1848. The City Sewers Act was a far-reaching piece of legislation; it vested all the sewers in the Commissioners, who were few in number (as opposed to fifty or so Commissioners of earlier Acts), all new houses built were to be connected to the drains, and old houses sited within fifty feet of a drain were compulsorily connected to it. The Commissioners also had full control over private cesspools (of which there were 5,400 in the square mile), and could compel each house to have a water storage cistern. The effectiveness of the Act and the exercise of its powers was ensured by the appointment of a part-time, but wholly effective, Medical Officer, Dr John Simon, whose extraordinary energy, far-sightedness, and skill in political manipulation within the City resulted in dramatic improvements.

Outside the City the Metropolitan Commission of Sewers had similar powers. However, due to dissensions within the Commission and in part because Chadwick, overburdened with work for the General Board of Health as well as the Commission, was no longer capable of overseeing the necessary improvements, the opportunity was wasted. The later career of the Metropolitan Commission of Sewers was dominated by bickering amongst the Commissioners and dissension with very little real progress towards unifying the sewers, or providing mains drainage, and London had to wait until 1855 for a new body, the Metropolitan Board of Works, before a similar opportunity was again presented.

However, cholera would not halt while sewer authorities quarrelled amongst themselves. The first outbreak of cholera in England was in Sunderland in 1831, following a gradual westward spread across Asia and Europe. Within a few months of the Sunderland outbreak it appeared also in Gateshead, Newcastle, Edinburgh, and Glasgow, where over 3,000 people died. Quarantine regulations were imposed in London on all shipping from northern ports, and on 6 February 1832 a day of national fasting and penance was announced. Four days later cholera appeared at Rotherhithe, from whence it travelled along the riverside to Limehouse, Lambeth, and Southwark, and later to the north of the City. Before the winter came and the outbreak declined some 5,300 people had died in London, and the next year during a less severe outbreak there were a further 1,500 fatal cases.

The causative organism in cholera is a comma-shaped bacillus, known as the cholera *vibrio*, which is ingested by the victim and lodges in the intestine. Its effects vary from case to case, sometimes the patient dies within hours of the first onset of pain, in others days of violent abdominal pain, vomiting, and diarrhoea may precede total collapse and death. In the 1840s one in every two cases was fatal. The disease is spread in two ways, by contact or by contaminated water. In overcrowded living conditions, which was the case for much of London's poorer working-class population, there are frequent opportunities for people to come into contact with the excrement of infected persons particularly when they are in the grip of the disease. Flies can also carry the *vibrio* from unclean clothing or bedding to food. The numerous cess-pits close to dwellings and open drains also clearly gave ample opportunity for the disease to spread. When spread by contact cholera thus tends to be confined to small areas, families, the tenants of a single house, or dwellers around a yard. However, the *vibrio* can survive for up to fourteen days in water, and as a waterborne disease cholera showed itself to have unique qualities of spreading from one area to another apparently capriciously, and, depending on the drinking water supply, striking at the well-to-do, the middle classes, and the labouring population indiscriminately. It was the very unpredictable nature of the disease that caused such terror in London and hastened much of the public health legislation of mid-nineteenth-century England.

The true nature of cholera was not known at this time, however, and indeed was not discovered until 1884 when the German bacteriologist Robert Koch, working in India, proved that the disease was caused by the *vibrio* which had to be swallowed to produce infection. Before that there were two conflicting schools of thought, one group believing that the disease, like so many others was spread by physical contact, the other holding that it was caused by the evil smells produced by rotting animal and vegetable matter, latent until activated by suitable weather conditions. The latter opinion, which was at the same time more naive yet perversely near to the truth, was held by several of the social reformers such as Edwin Chadwick and John Simon. Acting from their belief that evil miasmas caused disease, both saw a means of control in cleaning out cesspits, removing the piles of night-soil which the night men stored in their yards before its removal to the market gardens outside London, and providing adequate drainage. Legislation along these lines was enacted soon after the first arrival of cholera in London in 1832, and although powers lapsed after a short period it was the beginning of public health administration for the capital.

The greater powers of the 1848 Metropolitan Commission of Sewers and the City Sewers Act gave Chadwick and Simon respectively the opportunity to exorcise the miasmas of London. As we have seen, cesspits were closed, sewers were improved and house drains connected to them.

The result was no doubt an improvement, but had the most tragic results, for in 1849 cholera struck again in London but now the infected excrement was piped into the sewers, then regularly flushed by the sanitarians, and thus into the Thames. By August the disease was waterborne, and as the Thames was still a major supply of drinking water, the *vibrio* was carried far and wide throughout London. Deaths rose from 246 in June to 6,644 in September, and before the end of the year 14,000 Londoners had died.

The 1849 outbreak, however, produced one advance in knowledge. In this year Dr John Snow, later to achieve fame as a pioneer anaesthetist, who had a practice at Frith Street, Soho, published a pamphlet *On the Mode of Communication of Cholera*, in which he claimed that cholera was an intestinal disease and could be communicated through infected drinking water. The 1854 cholera outbreak gave him the opportunity to produce further evidence. In densely populated Soho over 500 people died within ten days within a 250-yard radius of the corner of Broad Street. Snow investigated and discovered that most of the affected families were obtaining their water from the pump at the corner of Broad Stıeet; his advice to the local Board of Guardians was to remove the handle of the pump (Rose, 1975) (some authors, e.g. Sheppard (1971) claim that the handle was chained) to prevent water being drawn. Snow continued his investigations in masterly fashion to the conclusion that the cause of the outbreak had been the disposal of water used for washing the napkins of a baby sick with cholera into a leaking cesspit only three feet away from the public pump.

Snow's work in the causation of cholera continued. He showed that the areas supplied by two south London water companies which were contiguous had strikingly different death rates from cholera in the 1854 outbreak. In one 13 per 1,000, in the other 3·7 per 1,000. As the conditions and social structure were much the same in both it seemed that these could have had little bearing on the death rates. As Snow pointed out, the Lambeth Water Company, which supplied water to the area with a low death rate, had its intake upstream of London at Thames Ditton, Surrey; the other company drew its water from the grossly polluted metropolitan reaches. Although the waterborne nature of cholera was not accepted by all, Snow's evidence, as well as visible pollution, produced the Metropolis Water Act of 1852 which prohibited the extraction of drinking water from the river below Teddington Weir.

The cholera epidemics, general disquiet about sanitation and the ineffectiveness of the Metropolitan Commission of Sewers eventually led to major reorganisation of the administration of London's public health, which came at a time of major remodelling of London's management in 1855 with the Metropolitan Management Act. Under the guidance of a most efficient metropolitan administrator, Sir Benjamin Hall (for whom the clock tower on the new Houses of Parliament was named Big Ben) the Metropolitan Board of Works was instituted. Its main function at first was to design and construct 'a system of sewerage which should prevent all or any part of the sewage within the metropolis from passing into the Thames in or near the metropolis' (Sheppard, 1971). Unfortunately, under the terms of the Act the Board was required to have parliamentary approval for all projects which were to cost more than £100,000 a condition which led to protracted delays and which was not repealed until 1858 – the year of the great stink – under the most pressing circumstances.

The intervening years (1855–8) were occupied by a succession of plans, counter-plans, and referees' reports with no real progress at the end. In 1856 the Board submitted three schemes for the construction of sewers which had been prepared by the Board's engineer Joseph Bazalgette (later Sir Joseph Bazalgette) to the Chief Commissioner of Works (Sir Benjamin Hall). The Chief Commissioner objected to parts of these plans and three independent referees (D. Galton, J. Simpson, and T. E. Blackwell) were appointed to consider the whole matter. The 1857 referees, as they were later called, made a careful study of the facts and commissioned two chemists to investigate the possible use of London's sewage as agricultural fertiliser after various forms of treatment. Happily, the 1857 referees eventually decided that irrigation with liquid manure would have been impractical because of the large areas of land that would be used, and abandoned this proposal, but they concluded that the only practical solution would be to build intercepting sewers to carry sewage eastwards, away from London. However, the Board objected to parts of the 1857 referees' report and referred them to their own engineer (Bazalgette) and two other civil engineers (G. P. Bidder and T. Hawksley) who in turn produced a counter-plan – which eventually was accepted. As is the way with such projects the costs had escalated from about £2·5 million for Bazalgette's original scheme, to £5·4 million for the 1857 referees' scheme, to a final estimate in 1858 of between £7 and £11 million!

However, while the engineers' plans were drawn up, modified or rejected, and then considered again, the condition of the Thames continually deteriorated. By now in hot weather the Thames stank unbearably.

The summer of 1858 was hot and the condition of the river became very obvious to the occupants of the Palace of Westminster. *The Times* of 3 July 1858 reported that three days earlier observers in one of the corridors of the Houses of Parliament,

> were suddenly surprised by the members of a committee rushing out of one of the rooms in the greatest haste and confusion . . . foremost among them being the Chancellor of the Exchequer, who, with a mass of papers in one hand and with his pocket handkerchief clutched in the other, and applied closely to his nose, with body half bent, hastened in dismay from the pestilential odour . . . other members of the committee also precipitately quitted the pestilential apartment, the disordered state of their papers, which they carried in their hands showing how imperatively they had received notice to quit.

The year 1858 represented in one way the climax of pollution in the Thames. When Queen Victoria and Prince Albert 'had attempted a short pleasure cruise on the Thames its malodorous waters drove them back to land within a few minutes' (Longford, 1971). The newspapers of the day gave the Thames prominent space, and Philip Howard (1975), in his excellent book *London's River*, reports that it excited more comment than the contemporary Indian Mutiny. Howard also quotes the statement of one Member of Parliament that, 'It is a notorious fact that Honourable Gentlemen sitting in the Committee Rooms and the library are utterly unable to remain there in consequence of the stench which arises from the river.' A partial solution was found in draping the windows of the ornate new building with curtains soaked in chloride of lime. There seems little doubt that the abominable condition of the river hastened the end of the restrictions under which the Metropolitan Board of Works laboured and they were now given a free hand to solve the problem of the Thames as quickly as possible.

The scheme finally adopted from the plans of Bazalgette, Bidder, and Hawksley was in essence to pipe the sewage away from London and discharge it well downstream. As most of the sewers and natural tributaries which were acting as sewers flowed into the Thames along a rough north-south axis, the flow could be intercepted by building west to east main sewers across London. The proposal was ambitious but brilliantly conceived, and Bazalgette as chief engineer of the Board deserves credit for what was one of the greatest achievements in a great age of engineering. His successful achievement was acknowledged by the Royal Commission on Metropolitan Sewage Discharge of 1884, but perhaps its most noteworthy aspect is that much of the sewage disposal system built by Bazalgette is in efficient use today.

On the north side of the Thames, three main sewers were built; a low-level sewer beginning at Chiswick and running close to the river to Abbey Mills, Stratford; a middle-level sewer which began at Kensal Green and ended at Abbey Mills; and the high-level sewer from Hampstead through Hackney, again terminating at Abbey Mills. From Abbey Mills the combined flow of the three sewers ran to the Northern Outfall at Barking (later the name Beckton was used for this region). The system was such that by following the configuration of the land the sewage in the high-level and middle-level sewers and from Abbey Mills to Barking flowed entirely by gravitation, only the low-level sewer required to be pumped up at Pimlico and at Abbey Mills. South of the Thames the levels of the land required a more complicated system. There were only two main sewers, the low-level running from Putney to Deptford and the high-level from Balham to Deptford. In both gravity caused the sewage to flow eastwards and from the junction at Deptford it continued by gravitation to Crossness on the Erith Marshes (the huge earth-covered rampart of the sewer can today be seen bisecting the site of the new town of Thamesmead, which is built partly on the site of the old Woolwich Arsenal). A smaller sewer was built across the low-lying areas of Bermondsey and Rotherhithe, and pumps raised the contents of both it and of the Effra River at Battersea into the low-level sewer. The only other pumping station required was where the low-level sewer joined the main outfall sewer at Deptford.

In all eighty-two miles of sewer were built or tunnelled, much of it through built-up areas which required underpinning of houses and entailed complicated planning and negotiation. Possibly more impressive than the size and complexity of the undertaking was the speed with which it was accomplished; from a start in 1858 both the major outfalls were functional in 1864, and the Prince of Wales formally opened the system in 1865. Bazalgette's achievement was to provide London with an adequate drainage system capable of carrying away most of the domestic sewage and the rainwater which fell in the London area. Unfortunately it did not cure the pollution of the Thames, although it undoubtedly removed the worst of the nuisance from the heart of London to deposit it firmly on its doorstep in Barking Reach. More positively, it provided the workable framework within which later improvements could be effected.

SEWAGE AND THE MIDDLE REACHES OF THE THAMES

The construction of the Northern and Southern Outfall sewers now meant that London's waste was taken downstream for discharge to the Thames. The sewers on the north were designed to carry a dry-weather flow of 72 million gallons per day (3·8 cubic metres per second) and rainwater of 178 million gallons per day (9·4 cubic metres per second), while those on the south carried a dry-weather flow of 36 million gallons per day (1·9 cubic metres per second) and rainwater of 108 million gallons per day (5·7 cubic metres per second). They discharged into reservoirs of nine and six acres respectively in which the sewage was stored. No treatment was given to the sewage, it was merely stored and discharged to the river from immediately after high tide for the first few hours of the ebb. Herein lay the fundamental flaw in Bazalgette's calculations; he claimed that this was equivalent to siting the sewers 12 miles (19 km) further downstream but the real distance seems to have been only 3 to 4 miles (4·8–6·4 km) (Water Pollution Research Laboratory, 1964). Because of the tidal nature of the river the sewage was trapped within the water-mass from Barking Reach downstream. The siting of the outfalls had been one of the objections to Bazalgette's earlier schemes and the 1857 referees had favoured outfalls further downstream in Sea Reach close to Canvey Island, a proposal that was rejected at least partially on the grounds of extra cost. With the advantage of hindsight it must be said that the referees' proposals were well founded and, had they been adopted, the lower Thames would probably have never suffered serious pollution in the twentieth century.

The discharge of untreated sewage in Barking Reach produced an immediate worsening of the river in this area. In May 1868 the Vicar and other senior inhabitants of Barking addressed a Memorial to the Home Office alleging that conditions in the river were a potential threat to the health of nearby communities, that fish were no longer to be found in the vicinity, and that the mud banks formed by the solid matter in the outfall discharges were a hazard to navigation. The resulting enquiry of 1869 held by Mr (later Sir) Robert Rawlinson produced a wealth of opinion in support of these contentions. However, Rawlinson concluded that there was no evidence that the polluted state of the river induced ill-health amongst the population of Barking. The absence of fish life was in his view not easily ascertained, an astonishing conclusion because there was abundant evidence that fish had been present in large numbers in Barking Creek and in the main river, before the outfalls were constructed.

This was a sensitive point in Barking which had a long history as a fishing port, the earliest reference being in 1320, until by the 1850s it was one of the greatest fishing ports in England employing 1,320 men and boys directly on the boats. Although the industry had declined by 1863 this was mainly due to social factors, especially the construction of railways which meant that the catch could be sent to the London market from ports, like Great Yarmouth, which were nearer the fishing grounds, and, indeed, by 1865 the large fishing fleet of the Hewetts was transferred from Barking to Great Yarmouth and Gorleston. Moreover, a severe storm in December 1863 off the Dutch coast caused much damage to the Barking fleet, and the deaths of 60 men, and was contributary to the demise of the fishing industry of the town (Powell, 1966). These factors apart it is not surprising that the inhabitants of Barking, with its strong tradition of fishing, should make much of the absence of fish due to pollution caused by London's sewage.

The problem of the mud banks in the middle reaches of the tidal Thames arose again directly after the Barking Enquiry when the Thames Conservancy (then responsible for navigation on the tidal Thames) attempted to prevent the Board of Works discharging untreated sewage. A decade of controversy between the Conservancy and the Board of Works ensued as to the degree by which the infamous mud banks of the area were caused by solid elements in the sewage. Eventually, an official enquiry was held during 1879 and 1880 which concluded that the shifting and increasing mud banks of the area were caused more by dredging navigation channels than by sewage discharge or natural siltation, a conclusion which it is difficult to accept today in the light of recent studies on siltation and in view of the heavy discharges of crude sewage to the river then made.

These enquiries and the more or less continuous complaints of the noxious state of the river from local authorities, shipping companies, and individuals led eventually to the establishment of the 1882 Royal Commission on Metropolitan Sewage Discharge under Lord Bramwell to 'enquire and report upon the system under which sewage is discharged into the Thames . . . , whether any evil effects result therefrom, and . . . what measures can be applied for remedying or preventing the same'. The evidence submitted to the Commission came from a wide range of lay persons and technical experts. The technical evidence was often contradictory, but had major importance in that it stimulated a great deal of research into the characteristics of the estuary and the effect of sewage in a tidal river. Thus, one witness demonstrated that chloride from sea water could be detected as far upstream as Chiswick, and clearly if sea water penetrated that far then so could sewage. The presence of an 'oxygen sag curve' in the river was demonstrated, water in the vicinity of the outfalls and downstream containing less dissolved oxygen than the water at Teddington or at the mouth of the river, and several experts pointed out that this was caused by the dissolved oxygen being used by the decay of organic matter. The correlation between poor oxygen conditions and the scarcity of fish was also pointed out. The Commission, in its second report in 1884, concluded that it was not justifiable to discharge crude sewage to the Thames, that it should be treated in some way so that liquid and solid elements were separated, the liquid discharged to the river on the ebb tide and the solid element dumped on the land or at sea. This set the rules for the future management of London's sewage, which broadly speaking are still in effect.

Much of the laymen's evidence to the Commission was of more personal interest and gives today an immediate picture of what conditions were like. Pilots of river craft and officers of the river police told of headaches and nausea caused by the foul odours given off by the water and mud, while the black, sticky mud gave off bubbles of bad-smelling gas. Fish which had formerly been brought up river alive in live-wells of boats now died in the polluted reaches, and no fish survived in many parts of the river where they had once been plentiful.

Just before the Commission was appointed, the condition of the river near the outfalls had been forcibly brought home to the public by the sinking of the *Princess Alice* on 3 September 1878. The *Princess Alice*, a paddle-driven pleasure ship on its return day-trip from Sheerness to London Bridge, was run into by the screw-steamer *Bywell Castle* in Gallions Reach (close to the Northern Sewage Outfall) and foundered. The *Princess Alice* was an unhandy ship overloaded with passengers who included a high proportion of family parties, women and children. Six hundred and fifty bodies were recovered from the river or found in the wreckage (Thurston, 1965). The tragedy evoked much comment in the national press at the time and there, and at the subsequent inquest, it was commented that the victims included several capable swimmers, who if unable to swim ashore had possibly been overcome by the noxious quality of the water in the river. Survivors averred at the inquest that, 'the water was very dreadful and nasty, it was in a very foul state indeed', and 'both the taste and the smell were something dreadful . . . it was the most horrid water I ever tasted and the smell was equally bad'. The Medical Officer of Health for the Port of London, a qualified medical man, had noted that the bodies recovered from the river had a kind of slime over them which one did not see on bodies in clean water and that 'the general condition of the river was that it smelt most horribly in the warm weather and was offensive in a variety of ways due to the northern and southern outfall sewers' (Thurston, 1965). Although the great Thames disaster was the worst civil shipping loss at the time in British history it did focus attention on the appalling condition of the Thames in the late nineteenth century and gave impetus to the creation of the Royal Commission on Sewage Discharges which was to lead to improvements in treatment.

An interesting contemporary report of observations on the effects of the polluted water on fishes was made by Dr Albert Günther, of the British Museum (Natural History), in what were probably the earliest experiments of their kind in Britain. Günther's report, dated 1883, to the Metropolitan Sewage Discharge Commission is quoted from his manuscript in the Library of the British Museum (Natural History). On 14 June 1882 Albert Günther wrote to Sir Joseph Bazalgette with an outline of his projected experiments and making some general observations.

> After having personally inspected the Sewage outfalls and the neighbouring parts of the river, I am prepared to submit to you the following observations:
> 1. Our experience of the effect of polluted water on the health or life of fishes is either of too general or too specific a nature to allow me to apply it to the conditions of the Thames existing at present, or to draw satisfactory conclusions from the chemical data contained in the papers, with which you have kindly supplied me.
> 2. Therefore, in my opinion, the question whether the water of the river, say within six miles from the outfalls, is so much polluted as to be injurious to the life of fish and crustaceans – can, at present, be answered by actual experiment only, viz. by mooring perforated fish-boxes with fish and crustaceans (shrimps) of suitable kinds, at various states of the tide and at different temperatures in different parts and depths of the river.
> 3. If the immunity of fish-life were demonstrated by these experiments, a second question would arise, viz. whether the water in this part of the river although not destructive to life, is yet sufficiently polluted to deter fishes from entering or passing through it; and whether this condition is permanent, periodical or

> only exceptional. This can be ascertained by the actual use of the net or trawl, and by enquiries made of fishermen. The latter kind of evidence may be taken as reliable, as far as it relates to the presence of certain fishes in the upper reaches of the river, and in the river below Gravesend. But as I am informed that fishermen have ceased to work, or nearly so, between Greenwich and Gravesend, their negative testimony must be tested by the use of the trawl or net.
>
> 4. There is no doubt that sewage contributes to a pollution of the water which may become dangerous to fishes, and that it contains (at least at times) very poisonous ingredients. But the proportion of this contribution as compared with that derived from other sources, is a question to be settled by the chemist rather than by the biologist; however, he may give some assistance by simultaneously instituting three sets of experiments, viz. by placing living fish,
>
> a. in sewage-water diluted by 2/5 (or any lesser proportion) parts of pure water.
>
> b. in sewage-water diluted by 2/5 (or any lesser proportion) parts of Thames water taken after the turn of the tide from mid-stream above the outfalls.
>
> c. in Thames-water taken at the same time and place.
>
> I have no great faith in the results of these experiments, which, unless continued for some time with due consideration of all incidental circumstances, must be rather uncertain, owing to the varying character of the sewage.
>
> 5. The lines of investigation thus indicated could not be satisfactorily followed out, unless at least a fortnight's continuous attention were given to the matter. Two experienced fishermen, with their boat (sailing-boat) seine and trawl, would have to be hired for that time. Of other apparatus, 10 fish-boxes, four large tubs, 6 small anchors and buoys for mooring the boxes, three thermometers would be required.

Günther evidently was able to devote some time to the fieldwork required through July and August 1882 and reported to the Commission in January 1883. His report is long and supported by full details of the experiments conducted by putting fish in floating and moored boxes in the river, and immersing others in buckets filled with river water and effluents. Some of his conclusions are quoted below:

> There are, at present, no fishes resident in the Reaches between Beckton and Purfleet, and some of the migratory fishes which are known to have gone up river as far as Woolwich or even Greenwich some 30 years ago, do not now pass beyond Gravesend. . . .
>
> But as other migratory fish, like Lampreys, Eels and Flounders are met with above London Bridge, it is evident that these fishes must be able to pass through the suspected district at certain times. . . .
>
> It can be experimentally proved that the water between Beckton and the vicinity of Purfleet is at certain points and at certain times unfit for fish-life, and that this unfitness consists chiefly in the absence of the requisite amount of oxygen in the water. . . .
>
> Sewage is not the only source of injurious pollution. The discharge from the Beckton Gas-works, though much less in quantity than sewage, makes up some measure by the intensity of its poisonous qualities, and I have no doubt that it materially contributes to the periodical unfitness of the water for fishes in the Gallion and Barking Reaches.

Günther's observations on the polluted state of the river and its effects on fishes also suggested that during the ebb tide the water became 'more and more unfit for fish-life', and this was especially true in summer when the area of 'deoxygenized' water was largest due to low fieshwater flow and high temperatures. His explanation for the presence of brackish-water fishes upstream of the outfalls was that lampreys (presumably the lampern, *Lampetra fluviatilis* was intended), flounders and probably eels all migrated between October and March when with low temperatures and high freshwater flows the river was in its less adverse state.

Some of Günther's experiments produced quite striking results. Thus of two lots of eels, flounders, and shrimps floated in the ebb tide in cages starting at Beckton, those at the surface survived better (one flounder was dead, the other sluggish after 70 minutes), than those sunk to a depth of 26 feet (when all the fish and shrimps were dead or dying after 35 minutes). In other experiments he placed eels (which are relatively tolerant of low oxygen conditions) in a three-gallon covered bucket of sewage, or sewage diluted with river water, and recorded the result. In 'pure' sewage the eels died within 30 to 45 minutes, but when diluted with thirty parts of river water the eels showed no sign of discomfort. A flounder, however, placed in sewage diluted with river water showed respiratory distress for the first hour, until the sediment had settled out, but survived the experiment. The toxicity of the effluent from Beckton Gas Works was also tested,

> an eel was placed into discharge from Beckton Gas Works. The effect was similar as if spirits of wine or some other highly irritant poison had been used. After a few violent struggles the fish succumbed, covered with a whitish mucus, and was dead in 5 minutes.

It might be added that Günther was very familiar with the effect of spirits of wine (70–80 per cent alcohol distilled from wine) as his collection of fishes at the British Museum was preserved in it (as it still is), and plunging a live fish into such alcohol is possibly the quickest way of killing it undamaged.

Dr Günther's general conclusions about the condition of the river and its fish life were presented in an appendix (E) of his report. Upstream of London Bridge the water was sufficiently pure for the ordinary freshwater fishes to live unharmed, and flounders were caught above Hammersmith. Above Barking and Beckton few fish or shrimps survived when caged in the water unless they were at the surface; the water was therefore toxic and no fish would be expected to occur at the worse times of the tide or in summer. At Crossness the water was of similar lethal quality. Downstream in the Long Reach, 30·6 km (19 miles) below London Bridge, and 9·7 km (6 miles) below the outfalls, the water was unfit for fishes at least during the later hours of the ebb tide, and not until Grays was reached, 4·8 km (3 miles) further on was the water at all fit for fish life.

These observations, which have been quoted at length because Albert Günther's work in this field has not previously been given the recognition it deserved, show the disastrous effects on the environment that discharging untreated sewage had. It also demonstrated the unfortunate effects of the decision to site the outfalls so close to London. The effluent from Beckton Gas Works was moreover indicted as a virulently toxic addition to a river which was already receiving an unbearable load of organic pollutants.

FROM THE ROYAL COMMISSION (1884) TO THE PIPPARD COMMITTEE (1961)

The Royal Commission in its second report on the disposal of London's sewage recommended that the discharge of untreated sewage should be discontinued. From 1884 various methods of precipitating the solid element from the liquid within the time available were tried. At first precipitation using lime and ferrous sulphate was attempted, while other experiments in deodorising the sewage with bleaching powder and sodium permanganate were made between 1884 and 1887 with uncertain results. The building of a series of sedimentation channels at the Northern Outfall Works began in 1887 and was completed in 1889, and from then until 1915 the lime and ferrous sulphate treatment was used to settle the sludge out of the sewage. At the Southern Outfall similar work was started in 1888 and completed in 1891 and treatment of the sewage was similar. The liquid resulting was discharged to the river, but the sludge, still very liquid and requiring 'dewatering' was taken by ship to the mouth of the estuary and dumped in the Barrow Deep, and after 1904 in the Black Deep.

The improvement to the river water appears not to have been conspicuous at first, although the authors of an 1891 report to the London County Council, which on its creation in 1891 had taken charge of the sewage outfall works, wrote that 'under average conditions there is little to reasonably complain of in the state of the river, but . . . during dry summer months and in particular places, the stream is still apt to become very discoloured, and occasionally to emit offensive odours' (Water Pollution Research Laboratory, 1964). By 1899 one of these authors, A. R. Binnie, in a further report to the LCC, strongly urged that further intercepting sewers should be constructed. They were needed partly because London's growth had continued and there was sewage to treat from a million inhabitants more than when the sewers had first been built, and because in conditions of heavy rain 'storm water' flooded the sewers and discharged direct to the river. These new sewers were begun in 1901. The combination of chemical precipitation and building additional sewers resulted in a striking return of fish to the reaches of the river which had previously been lifeless. Smelt and flounders returned to the river and were captured upstream in 1898–1900 (Cornish, 1902), whitebait returned to Gravesend and then Greenwich in 1892 and 1895 respectively (Water Pollution Research Laboratory 1964). The tidal Thames at the beginning of the twentieth century thus appeared to be recovering from the effects of more than half a century of gross pollution.

The improvement appears to have persisted approximately until the end of the First World War, and throughout this period and up to date there exists a mass of chemical information about the river, collected at first by the staff of the Metropolitan Board of Works, later by the London

County Council, the Greater London Council, and the Port of London Authority (which was the pollution control authority from 1909). This mass of information was condensed into the great *Effects of Polluting Discharges on the Thames Estuary* by the Water Pollution Research Laboratory (1964) and supplemented by further research by their own staff. There can be few, if any, other rivers in the world for which such meticulous chemical and physical records have been kept over such a period. Much of this information is too detailed to present here and the complexity of the situation is great indeed, the quality of the water varying with proximity to the outfalls, season, tidal cycle, rainfall, and freshwater flow over Teddington Weir. A simple parameter which illustrates the gradual decline in the condition of the river is the level of dissolved oxygen in the water over successive decades (Fig. 1).

In the early 1920s the condition of the water close to the outfalls had

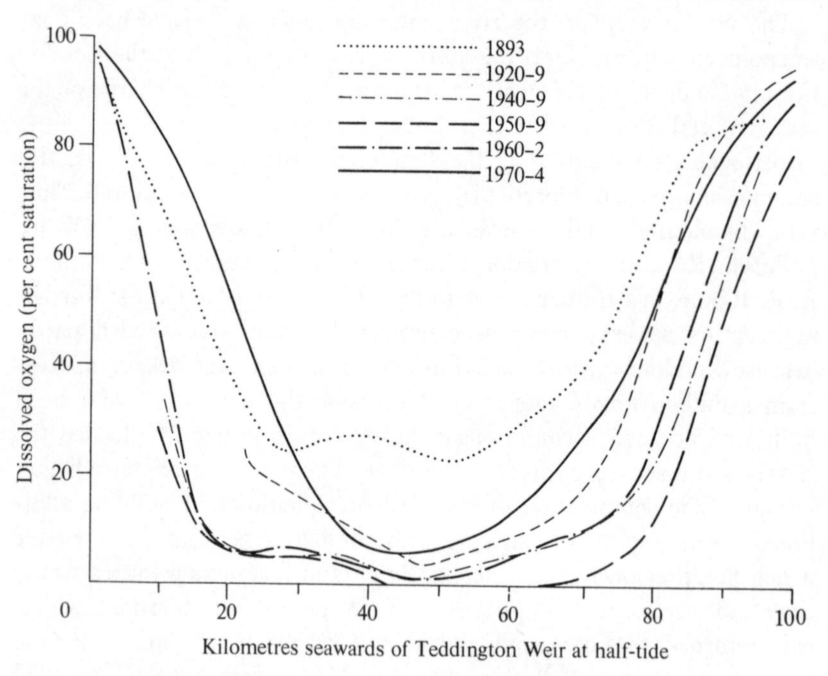

FIGURE 1 *Oxygen sag curves for July-September quarters with flows at 13 cubic metres per second at Teddington Weir*

deteriorated so that dissolved oxygen was only 5 per cent saturation in 1920, as opposed to 25 per cent in 1905 (Port of London Authority, 1967). In the immediately post-war period there was an upsurge in the population around London, but this was not balanced by a comparable extension to the sewage treatment works. This period also saw the development of much of the outer suburbs of London, which as they often lay outside the area covered by the main drainage of the LCC, were connected to small sewage works built about this period, sited on the tributary streams of the Thames. The result was that these small rivers now received sewage effluent often not of a high quality, so that pollution, to use a blanket term, spread widely through the natural waterways of the London area.

However, the inter-war decades were not entirely unproductive. A pilot scheme at the Northern Outfall Works to provide an activated sludge plant to give secondary treatment to 5 million gallons of sewage per day (0·26 cubic metres per second) was commenced in 1929 and opened in 1931, and was expanded in 1933 to treat 30 million gallons per day (1·58 cubic metres per second). This secondary treatment resulted in an improvement in the quality of the effluent discharged to the river from Beckton, the larger of the two great outfalls. But the volume of sewage reaching the outfalls continued to increase, and the poor-quality effluent from the many local sewage treatment plants discharging directly, or indirectly, to the Thames more than compensated for the improvement, and the general trend was a worsening of the river.

The condition of the river and of sewage treatment in general was once again of major concern. The Report on Greater London Drainage by R. J. Taylor, Sir George W. Humphreys, and T. Pierson Frank to the Minister of Health in 1935 (Port of London Authority, 1967) recommended the consolidation of all arrangements for sewage disposal within the area and the reduction in number of sewage treatment works from nearly two hundred to ten large regional works. The Middlesex County Council had been planning along these lines and in December 1935 commissioned a large regional sewage treatment plant at Mogden, which allowed no fewer than twenty-eight small, often inefficient, local authority works to be closed down. This far-sighted example of the advantages of centralised treatment was not followed elsewhere because all such proposals came to an end with the outbreak of the Second World War. By the end of the war the river was again in a severely polluted state. Lack of funds for new plant, bombing, and worn out existing plant meant that sewage effluents discharged to the river were often of a low standard.

At this time anaerobic (absence of dissolved oxygen in the water) conditions were virtually continuous throughout the year in the vicinity of the outfalls. Hydrogen sulphide was liberated in large quantities, and for many miles of the lower river the water literally stank. It is said that, 'silver in ships' saloons and buttons on uniform jackets were tarnished by sulphurous fumes from the filthy water. Suicides from London's bridges were not so much drowned as poisoned, and stomach pumps were part of standard rescue equipment' (Bates 1977). There was said to exist in the Crossness Sewage Works' archives a very substantial file of complaints by ships' masters and shipping companies concerning the tarnishing of the brasswork, carefully polished for entry to the Port of London but fouled by the gases from the water on the passage upriver. From personal experience I can say that in 1949, when I was stationed at Woolwich and living in Essex, on hot summer's evenings it was often desirable when on one's way home to walk under the Thames by the pedestrian tunnel rather than endure the smell stirred up by the churning paddles of the Woolwich ferry boats. Nor was it necessary to look out of the window as one approached North Woolwich by train or trolley bus, the smell of the river gave an accurate fix on one's progress!

Chemically the water could not have been much worse. The year 1949 was very dry and the upland flow into the head of the estuary was low and as a result the river was in particularly bad condition. Even in the first quarter of 1950, when with the rainfall of January to March and lower temperatures the level of dissolved oxygen should have been high, there were nearly 19·3 km (12 miles) of river below 5 per cent saturation, and in the third, and expectedly the worst, quarter the 5 per cent level or below extended for 56·3 km (35 miles), and for 19·3 km (12 miles) there was no oxygen detectable.

The Thames Survey Committee was set up in 1949 by the Water Pollution Research Board at the instigation of the Port of London Authority to investigate the causes of pollution and siltation in the river. At this time it was felt that the heated effluents discharged by the numerous Thames-side electricity generating stations had an adverse effect on the condition of the water (indeed between 1920 and 1960 the temperature of the water had risen by 3·3°C, for this and other reasons) and the Pippard Committee (named for Professor A. J. S. Pippard, its chairman) was appointed to examine this problem with the Thames Survey Committee. Both groups found that the situation was so complex that a considerable amount of basic research was required, and many old theories scrutinised, to arrive at any reasonable solution. The Pippard Committee's report appeared in 1961 (Pippard Committee, 1961) after close liaison with the Thames Survey Committee which provided much of the background information, and whose own report was not published until 1964. Meanwhile, while the Committee had been gathering data and deliberating, the London County Council, at the instigation of Professor Pippard had been allowed by central government to improve the sewage treatment plants at the outfalls. At Beckton Outfall works a second activated sludge plant to treat a further 70 million gallons per day (3·7 cubic metres per second), thus giving it the capacity of secondary treatment to half of its normal dry-weather flow was opened in October 1959, while a full treatment plant was built at Crossness Outfall Works by 1964.

The Pippard Committee report recognised the prime role of dissolved oxygen in the purification of the polluting loads in the estuary, and set the objective to be attained as the presence of a reserve of dissolved oxygen sufficient to deal with accidental discharges, storm-water overflows, or other crises, at all times and in all places (Tetlow, 1971). The main conclusions of the Committee were that: most of the polluting load came from the five large sewage works at Mogden, Beckton, Crossness, Dagenham, and West Kent and that further improvements at all works would be required; industrial pollutants amounted to 10 per cent of the total and were mainly discharged in the lower reaches; power-station effluents had an effect on the condition of the water, as raised temperatures accelerated bacterial activity and depressed the oxygen-holding capacity of water. In addition, the flue gases from two central London power stations were washed (to reduce atmospheric pollution) and the resulting discharges decreased the level of dissolved oxygen further.

Several other proposals to improve the quality of the river and for its future management were also produced. The Thames Survey Committee report, backed by its wealth of detail, provided the technical background for this management. One of the points that was made by both committees were the quantification of the level of dissolved oxygen required if salmon were again to migrate into the Thames. The critical period was felt to be April and May when smolts, young salmon, are migrating to the sea, and these were known to require a level of dissolved oxygen of not less than 30 per cent of saturation. The Thames Survey Committee concluded, 'it might well be that a salmon fishery could be established if, during these two months, the minimum concentration of dissolved oxygen in the estuary were not lower than 30 per cent of the saturation value in nine years out of ten.' It was recognised there that the regular passage of migrating salmon would represent the crowning achievement in purifying the river, which at that time was in danger of posing a threat to public health.

THE CLEANER RIVER THAMES

Following the reports of these two Committees a determined and well-planned effort to put their recommendations into effect was made by the Port of London Authority (the pollution control authority) and the London County Council, and its successor, the Greater London Council. The improvements made at the two outfall works before 1964 have already been mentioned. The improvement to the condition of the river water was very quickly reflected by the detection of the presence of dissolved oxygen throughout the year. From 1966 no anaerobic reach existed in the Thames (Gameson and Wheeler, 1977), although often, particularly during the summer and autumn, the level of dissolved oxygen was perilously low. Positive action by the authorities was still being taken; Beckton Sewage Treatment works was virtually rebuilt in the 1970s so that it was capacious enough to give full treatment to the whole of the sewage flow, with reserve capacity to allow for increased flows over the next twenty-five years. As with the rebuilt Crossness Works, the opportunity was taken to close down some of the smaller, out-of-date works and pipe the sewage they had formerly treated into the major works. As many of these works had discharged effluent to Thames tributaries these began to reflect the general improvement to the main river.

The Mogden Sewage Treatment works was extended between 1956 and 1962, and a source of pollution in the upper reaches of the tidal Thames thereby removed. By 1973 the effluent discharged to the river was of a high quality and the effect of this works was minimal.

More recently, the Long Reach (West Kent) works has received attention. Formerly, its contribution to the pollution of the river was negligible by comparision with the other sewage works. However, as these were improved the significance of this works increased, and its flow has grown progressively over the years. This treatment works was being rebuilt in 1977 so that the polluting load will in future be decreased.

Other sources of pollution have been reduced. In some instances industrial concerns have installed or improved treatment plants of their own, in others the effluent from industrial processes is taken into the sewers or removed for special treatment. This has resulted in the opportunity to monitor the discharges of dangerous contaminants such as cadmium, mercury, polychlorinated biphenyls, and pesticides, and to deal with any excess discharge before they can build up in the food chains of the estuary. In one case, at the Thames Board Mills at Purfleet, the problem of the polluting discharge of 5 million gallons per day which takes a considerable volume of oxygen from the river, was amply compensated for by the company's installing two huge aerators in the Thames at Erith. This plant raises the level of dissolved oxygen in the river at a point where it has the most beneficial effect.

The discharge of heated water to the river mostly by electricity generating stations has changed also, but this is as much due to developments within the power industry as to pollution control. Since 1960 new power stations have been built downstream, where the greater volume of water can absorb the excess heat with less serious effects, and many of the upstream stations have been closed down, or are operated only at peak demand periods. Whatever the causes, the effect has been that the temperature rise in the river within London caused by thermal pollution has been less than in the previous decades, which must have a beneficial effect on the condition of the river and its native wildlife.

The electricity industry has, however, been enabled to make a further contribution to the greater cleanliness of the Thames in that from 1970 it was no longer required to wash the flue gases from Battersea B power station. When the station was first built in 1929 it was subject to the condition that the flue gas was washed mainly so as to remove sulphur. Experience later showed that the cool, washed gas often fell rapidly to the ground to create an unpleasant mist and smell locally, while the washings containing sulphur dioxide and other toxic material was poured into the Thames. This original planning decision, the wrong decision for the right motives, was later described by the distinguished chemical engineer Dr David Train, who acted as chemical consultant to the Port of London Authority, as 'a grave technical error'. It certainly resulted in an additional polluting load to the Thames without resulting in a compensating atmospheric environmental improvement.

Improvements of all kinds have been the feature all along the tidal Thames, and throughout the rivers of the Greater London Council area for the two decades from 1960. The restoration of the Thames Estuary has been achieved by the devoted work and unsurpassed skills of individuals and organisations, amongst whom the Port of London Authority, the Greater London Council, and its predecessor the London County Council are pre-eminent.

The restoration of London's river and the re-establishment of its wildlife is a feat of unsurpassed credit which at the time it was achieved was unique for its scale in the world.

THREE

FISH AND FISHERIES OF THE TIDAL THAMES

SOURCES

Most discussions of marine and estuarine fisheries today can be based on the amassed fishery statistics which have been collected over some ninety years from local fishery offices and other sources. Although the Scottish fisheries for herring were statistically recorded from 1809, the regular collection of figures for weights landed in England and Wales was not attempted until 1885, two years after HRH the Duke of Saxe-Coburg, then Duke of Edinburgh, collected through the Coastguard some figures for landings for a paper read at a conference at the International Sea-Fisheries Exhibition in London (Burgess, 1967). Although fundamental in recording the variations in the fishing industry and possibly shedding some light on the fluctuations in the abundance of food fishes, these figures are too general and their collection started too late to throw any light on the abundance of fish in the tidal Thames. Such sources as are available are not statistical in the modern sense, but together they give a general impression of the size and value of the fisheries and of the status of the fishes and shellfish which supported them.

Two of the major sources can be mentioned here. Foremost among them is William Yarrell (1784–1856), a London stationer and newsagent and honorary officer in various capacities of the Zoological Society of London and of the Linnean Society of London, whose books on British birds and fishes were of fundamental importance in promoting the study of natural history in England. Yarrell's *A History of British Fishes* was first published in 1836 and the third edition appeared posthumously in 1859. The importance of Yarrell's notes in his book and serial publications is that, being a Londoner, he had a personal knowledge of the fishes in the Thames, and in his time he saw the virtual destruction of the fisheries of the upstream tidal river. A second major source is the published and manuscript writings of James Murie (1832–1925), who curiously was also associated with both the above London societies, although as a member of their salaried staffs. Murie, a Scot, qualified in medicine at Glasgow University and then took successive posts as assistant in the Royal College of Surgeons' Museum, ship's surgeon, and medical officer on John Petherick's expedition towards the sources of the Nile, later becoming prosector at the London Zoo, and Librarian at the Linnean Society, and eventually retired to Leigh-on-Sea, on the Essex side of the estuary (Mitchell, 1929). Here, as a member of the Kent and Essex Sea Fisheries Committee, he began to write the *Thames Estuary Sea Fisheries*. The first volume was published in 1904, the remainder was found incomplete and still in manuscript after his death in 1925. It was copied in manuscript and the illustrations were saved by Mr W. Pollitt, then borough librarian at Southend-on-Sea, in the Reference Library of which borough the copy is still preserved. Murie's work contains a vast accumulation of knowledge of the outer estuary fisheries and fish from the second half of the nineteenth century, for not only did he record his own observations, he evidently established cordial relations with the local fishermen (something he failed to do with his scientific peers in London) and was given the advantage of their practical knowledge.

Two other sources of information are less satisfactory, but nevertheless important in establishing the status of fish in the Thames at an earlier date. Robert Binnell's *A Description of the River Thames* ... (1758) is a mine of information on its main theme, the fishery regulations made by the Conservators of the river (at that time the City of London), but Binnell's natural history was weak and his assessment of the value of its fishery suspect (for, as water-bailiff, he was responsible for their maintenance). The second is a topographical work, *Some Account of London* (1791), but topography with a difference, for its author was the naturalist Thomas Pennant (1726–98), whose *British Zoology* (numerous editions between 1766 and 1812) had been the fundamental English-language source of knowledge of animals through the late eighteenth and early nineteenth centuries.

MIGRATORY FISH AND FISHERIES

Although evidence is sparse before the eighteenth century, it is probably safe to assume that from its earliest days London's inhabitants derived much of their food fish from the tidal river. Smelt bones have been identified in several layers dating from Roman Southwark and medieval Westminster, and sturgeon scutes in medieval remains from the Westminster Abbey site and Tudor Baynard's Castle (where salmon bones have also been identified). All are species which could have been caught in the adjacent river.

The fishery for smelt was probably the most important of all tidal Thames fisheries. The smelt is migratory, living in the outer estuary of rivers, entering the river mouth in winter (December to February) and moving upstream to spawn in March to May in the region of the river where tidal influence is strong but the water is of low salinity. Smelt-fishing must therefore be seasonal to some extent, the largest catches being made in winter and early spring when the adults are migrating. The smelt is a delicious food-fish with a delicate flavour; when fresh it smells strongly of cucumber. It grows to a length of 250 mm (10 ins).

In 1630 Sir Robert Ducie, Lord Mayor of London and Conservator of the River Thames, forbade smelt fishing westwards of London to Isleworth church from 10 March until 14 September (thus protecting the spawning stock), and to the east of London from 21 October until the Good Friday following. Trink nets which were used for smelts had to have a mesh of 'full two inches' in the forepart, narrowing towards the cod-end to 1¼ inches. In Robert Binnell's time (*c.* 1750) smelt were caught at Richmond in March, but later in the summer they were most abundant around Blackwall and Greenwich.

I am much indebted to Mr C. Kirke Swann for copies of notes inserted in a copy of Pennant's *British Zoology* 4th edition (1776–77), which belonged to Sir Richard Gregory (1766–1840), referring to the Thames smelt fishery between 1797 and 1828. These notes take the form of newspaper clippings (source not noted) and two quoted petitions from the body of Fishermen on the River Thames to the Lord Mayor at Mansion-House. The note dated 4 April 1797 reads,

> Smelts have been so plentiful in the River lately, that on Wednesday the fishermen disposed of them on the Banks of the Thames at the rate of 2d. a basket full, containing near one hundred; and on Monday, in Deptford Creek, the draught was so great that they were sold in the manner of sprats, by coal measure. . . .

> (24 February 1798) Mansion-House, Saturday. – The season of the year being uncommonly forward, and, in consequence, the *Smelts*, which are seldom fit to take till the beginning of April, being now in the Thames in large quantities, several persons, deputed by the main body of Fishermen, applied to the Lord Mayor . . . to allow them to begin fishing for Smelts immediately, in place of waiting till the 25th of March, as fixed by law. His Lordship took till Tuesday next to consider their request; which, if granted, will employ near 500 persons, and furnish a liberal supply to the markets.

Forty years later Yarrell reported a decrease in numbers,

> formerly, the Thames from Wandsworth to Putney Bridge, and from thence upwards to . . . the bridge of Hammersmith, produced abundance of smelts, and from thirty to forty boats might then be seen working together, but very few are now to be taken, the state of the water, it is believed, preventing the fish advancing so high up.

However, the state of the water was only one of several factors affecting this fish. In April 1828 the City of London water-bailiff introduced a deputation from the body of Fishermen on the River Thames claiming that the whitebait fishery prosecuted by 'certain persons' with unlawful nets 'had totally destroyed the fishery on the River, and had, in consequence, reduced the petitioners and their wives and families to a state of starvation.' Evidence was given that some years earlier smelts were sold in the market at from 1 shilling and sixpence to 2 shillings per hundred but that now 'they could scarcely be got at any price'.

At the hearings of the Commissioners of Salmon Fisheries in January 1861 (Salmon Fisheries Commission, 1861), Mr Henry Farnell, then Secretary of the Thames Angling Preservation Society, said that the smelt gradually declined in numbers but that they came up the river 'if there was a good flush of water'. Perhaps in 1860 there had been high flows of water, for in that year they were caught in numbers. One fisherman, Lewis Gibson of Putney, received 16 shillings a hundred, and one lot (presumably a day's catch) earned him £10. But as Farnell observed, he had not heard of catches of this size for fully thirty years. By the end of the nineteenth century smelt were only commercially fished in the mouth of the river. Murie claimed that they congregated along the Blyth Sands from the Lower Hope Point to the Grain Spit and during winter and spring the Southend boats fished here for them regularly and 'derive a considerable income therefrom'. He also found that smelt were to be encountered in most of the creeks and bays between Yantlet and Mucking Flats, although presumably they were not fished for in these regions. That the smelt had declined in numbers was suggested by the evidence of Ebenezer Newby, a fish salesman at Billingsgate market, to the Commissioners for Sea Fisheries in 1878, who said, 'There was a time when there were smelts enough in the Thames to supply London,' but that 'the smelts have fallen off like everything else . . .'. At this period, as in Murie's time twenty years later, smelt fishing was prosecuted in the mouth of the estuary, the fishery in the Medway still being particularly valuable. The Foreman of the Fishery of the city of Rochester told the Commissioners that smelt was the principal fishery between Rochester and Sheerness and that the fish were increasing in numbers. The Medway smelts continued to support a fishery in the 1920s and 1930s, although declining in numbers (Marsh, 1971), and by the end of the 1950s they had left the river (Waters, 1964).

In the tidal Thames itself, as elsewhere, smelt varied in numbers from year to year. In natural conditions this is caused mainly by exceptionally good survival of the larvae and fry producing a strong 'year class' which leads to more numerous fish in the river and in time greater catches, lasting until mortality caused by natural deaths or fishing exhausts the year class. In the Thames this natural, but irregular, fluctuation may have been supplemented by changes caused by temporary improvement in the condition of the river (in some cases caused by improved sewage treatment, in others by increased freshwater flow after heavy rainfall). Thus, as we have seen, in 1798 they were abundant, 1860 saw many smelt in the upstream river, and others were reported in 1868 (Kew Bridge and Teddington), 1898 (Richmond), 1899 (Westminster) in 1900 beyond Blackwall (Murie, 1903) and at Putney in August, and a month later they were caught in numbers at Isleworth, Kew, and Teddington (Cornish, 1902).

The salmon was another Thames fish which, because of the high price at which it sold, was important, although it is doubtful whether it contributed much in terms of weight landed. It is difficult to evaluate the comparative abundance of the salmon in the Thames, partly because the 'king of fishes' tends to be recorded far more than other fishes and because of the appeal to the popular imagination that it commands. Examples of the latter can be cited. The capture of a salmon in the Thames in 1974 created an intense interest in the national press, radio, and television, and yet it was only a single fish of minor importance to the ecology of a river in which thousands of specimens of other kinds of fish had been abundant for several years without attracting much attention. Second, the widespread belief, cited by so many authors, that salmon were so numerous in the Thames that apprentices' indentures specified that they were not to be fed on salmon more than once a week. This is a myth, but its widespread belief serves to show how any figures about the salmon tend toward exaggeration. The salmon and apprentices story has been fully examined by Cohen (1955) and his careful analysis of the so-called evidence failed to reveal any indenture with the salmon clause in it. The myth is current of Edinburgh, Newcastle, as well as London apprentices' indentures. I myself have invited authors making the statement that London apprentices' indentures included the salmon clause to produce their evidence but in every case the authority cited was another book, or 'general tradition', or my 'grandfather told me he had seen it', which is exactly the response Mr Cohen elicited. I share with Mr Cohen, whose lucid study should be consulted, the belief that there never was such a clause in apprentices' indentures.

That the salmon was found commonly in the Thames at one time is certain. That it was especially abundant is unlikely. In the first place, the Thames as a whole is a lowland river, in which even in its natural state suitable spawning grounds for the adults and nursery grounds for the young would be sparse when compared with rivers such as the Severn, the Wye or the Usk which rise in mountainous regions. Second, the Thames, like the East Anglian rivers, had a richer indigenous freshwater fish fauna on account of the post-glacial connecting land-bridge to Europe across the present-day North Sea. This richer fauna would have resulted in the Thames salmon encountering more numerous competitors, such as some members of the carp family and predators such as pike and perch, than in western or northern rivers (Wheeler, 1977). There is thus reason to suspect that the Thames was not so prolific a salmon river as other large rivers in the British Isles; a suspicion that is reinforced by records of the import of salmon to Billingsgate from the Lake District in the seventeenth century (Clarkson, 1971) and in the early eighteenth century (Crofts, 1967). If it was necessary to import salmon into London to satisfy the demand at that period, when the native stock was still present, it must be assumed that the salmon was not so abundant as has been made out by some uncritical authors.

Despite this, the salmon did form a valuable fishery in the Thames. The regulations issued by the City of London for the conservancy of the river below Staines Stone mention salmon regularly. Thus in 1630 fishermen were forbidden to shoot salmon nets after 14 September because 'they are out of season, and remain here upon the River only to spawn and breed'. In 1741 it was ordered that only salmon nets with mesh larger than three inches should be used from Kew westwards to the City Stone at Staines bridge, from 10 March to 14 September (Binnell, 1758) the object being to allow small and young salmon to escape. Regulations of this kind, often with variations in the dates and other details, seem to have been fairly regularly issued through to the nineteenth century.

Thomas Pennant (1791) was quite matter of fact about the salmon in the tidal Thames; he wrote that 'it appears in the river about the middle of February, is in great estimation, and sells at a vast price'. Fishing stations existed then in what is today the heart of London. The Wandsworth church-wardens' book for 1580 records, 'In the somer, the fysshers of wandsworth tooke between Monday and Saturday, seven score salmon . . .'. At Chelsea the fishery was financed by the Lord of the Manor and two others and during one week of May 1664, nine salmon were taken, with a total weight of 172½ lb. This fishery extended virtually from Battersea to Lambeth. Fulham was also a fishing village and continued so in a small way to 1813, although only one salmon was captured that year (Williams, 1946). It is also claimed that 130 salmon were sent to market on one day in 1766 (Williams, 1946; Water Pollution Research Laboratory, 1964). By the time William Yarrell compiled his *A History of British Fishes* (1836), however, the salmon had become scarce. He wrote, 'A Thames Salmon is a prize to a fisherman, which, like other prizes, occurs but seldom. The last Thames Salmon I have a note of was taken in June 1833.' Yarrell was fortunate, in a sense, for as a young man he had seen Thames salmon, and no doubt caught them by

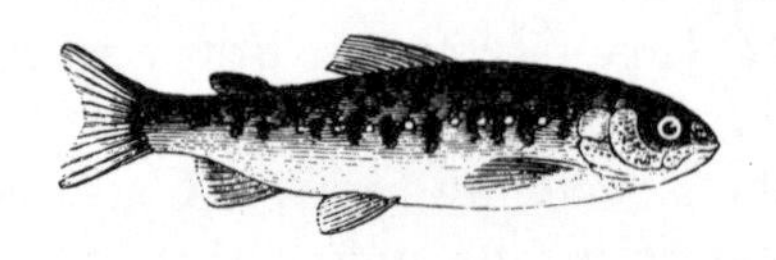

FIGURE 2 *Fry of Thames salmon*

angling; he also illustrated the fry from a Thames specimen, for they were 'occasionally caught in the season . . . by fishermen who work at night with a casting-net on the gravelly shallows for Gudgeons . . .' although what he called a 'fry' would today be known as a salmon smolt (that is, it had become silvery overall and was on the verge of migrating to the sea).

The scientific nomenclature of the salmon family caused considerable problems to nineteenth-century naturalists, for not only were many of the local races (such as lake trout and sea trout) regarded as distinct species but the young of the salmon was thought to be another distinct species. Yarrell (1836) followed his forerunners in recognising the parr or samlet as a separate species – he named it *Salmo samulus*, although later (Yarrell, 1859) he accepted the parr as a juvenile stage of the salmon or the trout. While parr has become an accepted general English term for the young of the salmonid fishes which bear a series of rounded blackish smudges on the sides – known as 'parr-marks', in the nineteenth century many regions had local names for them, heppar in Devonshire, branlin or samlet near Carlisle, and skegger on the Thames. Skeggers (also spelled skeggars) had been abundant before the 1830s according to Yarrell, but had declined in numbers. He wrote,

> Laleham, between Staines and Chertsey, where the water is shallow, formerly afforded the greatest quantity; forty or even fifty dozen have been taken in one day by a skilful fly-fisher; but the numerous gas and other manufactories on the banks of the river are considered so greatly to have affected the quality of the water, that a Salmon or a Skegger in the Thames is now but rarely seen.

Despite the confusion in nomenclature that Yarrell and other writers suffered, at least one contemporary fisherman on the Thames knew that skeggers were young salmon. Thomas Milburn of Hampton who described himself in 1861 (Salmon Fisheries Commission, 1861) as a fisherman living within the city jurisdiction and who had fished the river for seventy years (he was then nearly 80 years of age), recognised two sorts of skeggers 'one was a little bright fish . . . the other . . . is a marked spotted fish'. Possibly one was a salmon parr, the other trout parr. Milburn had taken gentlemen anglers to Windsor who caught '18 or 20 dozen skeggers a day; one half of them would be small salmon'. Henry Farnell, then Secretary of the Thames Angling Preservation Society, who lived at Isleworth also had boyhood memories of catching skeggers. 'I recollect taking them at Chertsey Bridge . . . we used to catch them when we were fishing for gudgeons . . . eight to ten dozen a day'. The distinguished naturalist and artist John Gould, FRS, also recalled his boyhood memories of 1822 or 1823 when salmon fry (presumably parr) were caught in great abundance then and prior to that time. He claimed the skeggers were most common at Laleham, of which, Samuel Harris, a fisherman who worked between Staines and Chertsey Bridge, also had boyhood memories. Harris's brother used to 'mind the ferry at Laleham' and on his numerous crossings of the river trailed two lines each with two flies and a small maggot on the hook over the stern to catch 60 or 70 skeggers a day. Laleham 'was the grandest piece of water you could find in the river for them' according to this witness (Salmon Fisheries Commission, 1861).

It was significant that all these witnesses, gentlemen anglers, Fellows of the Royal Society, and unlettered professional fishermen, had to recall the days of their boyhood to discuss skeggers in the Thames. One by one they gave evidence to the Commissioners for Salmon Fisheries at their hearing in London in January 1861, all to say that young salmon were no longer found in the Thames. With the extinction of the young fish the stock clearly could not survive long and with it would pass from English usage the peculiarly Thames-side term for parr – skegger.

Salmon by this time were virtually extinct in the whole Thames system, and their passing was recorded both at this enquiry and elsewhere, and because of the interest salmon possesses it is proper to give it in detail.

John Gould recalled seeing adult salmon at the tails of the weirs, especially at Windsor in 1822 or 1823; he believed these were salmon kelts, fish which had spawned and were drifting downstream in an exhausted condition. But, he reported, the bargemen about this season used to capture large numbers of these fish – they called them 'strikes' – and 'they were killed without mercy or thought for the future . . .'. Kelts when caught by the professional Thames fishermen had little value, but even so they were kept. Thomas Milburn, who had fished at Hampton from about 1790, remembered catching four or five in the course of the spring, some of which weighed up to 15 or 16 lb, but as he observed, 'they were of no use, and we used to give them to the poor people to eat. They were never fit to go to any other people'.

Milburn must have been one of the last professional Thames fishermen who had made a substantial part of his living from the salmon. Although he kept no strict account, he estimated that he caught between eighteen and twenty a year in his draft net – but he 'did not work so hard for salmon. I used to attend upon gentlemen angling.' His memory (which ranged back to the 1790s) was that they generally started fishing for salmon in March and continued through the autumn when most of the smaller fish of 10 to 12 lb weight (which they knew as harvest cocks) were caught. When salmon fresh from the sea were caught early in the season they would fetch 12 shillings a pound, and 10 shillings a pound was a regular price. However in times of heavy catches the price might drop to 1 shilling a pound. The biggest salmon Milburn caught was a 41 lb fish, for which he received 7 shillings a pound.

Many other fishermen were catching salmon at this period. Within the City's jurisdiction, that is downstream from Staines, there were local fishing communities at Sunbury and Isleworth, the latter being the most important of all on the upper tidal Thames. Judging from the catches made by Milburn at Hampton, and from his estimate of a hundred fishermen at Rails Head, Isleworth, the annual catch from this region must have been of several hundred fish. Indeed, Williams (1946) credits the Thames fishermen with catching up to 3,000 salmon in a good year as late as 1810, a figure which seems startingly high and for which no evidence is presented. Mr Williams gives other figures which are of the greatest interest and sound more probable; forty-seven fish caught at Richmond Bridge on 7 June 1749, ten salmon and three thousand smelts caught at Wandsworth in 1810, and he states that there were over four hundred professional fishermen working on the Thames in 1798.

The River Lee (known in the Waltham Abbey–Chingford area as Lea), which was the largest natural tributary of the tidal Thames, also contained salmon, although judging from the sparsity of records probably not a large stock. Farmer (1753) reported that some salmon were captured in the river there (approximately 12 miles from its confluence with the Thames). According to Brian Waters's delightful *Thirteen Rivers to the Thames* (1964) the Cornmill Stream, just behind the abbey which the last English king, Harold, founded in 1060, contained a salmon leap in the 1800s. Mr Waters also quotes Sir Henry Chauncey's *The Historical Antiquities of Hertfordshire*, that 'if the salmon had free passage by the mills and through the sluices at Waltham and were preserved from poachers they would greatly increase in the river, and could be a great benefit as well to the city of London as the county'. J. E. Harting (1894), in an article concerned with Izaak Walton and the River Lea, reported from literary records the capture in January 1816 of a 28 lb salmon at Lea Bridge (Hackney), two fish (6¾ lb and 4½ lb) a little above Old Ford (near Bow) in April 1825, and a 15 lb salmon at Walthamstow Ferry on 9 December 1833. The year 1816 was said to be the greatest salmon season ever known in the tidal Thames (it was also an extremely wet year) – and ninety salmon were caught in one morning in the lower

Thames. In the River Lee that year, the miller at the mills near Hackney (presumably using the fish traps that mills are uniquely placed to employ) captured about 670 lb of salmon. The abundance of salmon that year depressed the price, and at Billingsgate it was possible to buy salmon at 3 pence a pound (Salmon Fisheries Commission, 1861).

One of the most .fascinating documents concerning salmon in the Thames is the extract of a journal kept by two fishermen at Boulter's Lock, near Taplow Mills, Maidenhead. The Lovegroves, father and son, kept a record of all the salmon taken at Boulter's Lock between 1794 and 1821, and extracts from this were presented to the Commissioners for Salmon Fisheries in January 1861 by Mr Richard Lovegrove, the grandson of the original journal writer. The original notes were also made available to Venables (1874) who gave full details from them. The Boulter's Lock figures have been summarised many times before but in their entirety they represent the nearest approach that we have to a statistical record for salmon in the Thames. The important details are given in Table 3.1.

The Lovegroves' journal was rich in ancillary notes. The price that their salmon fetched was occasionally entered; two fish of 18 and 21 lb brought 5 shillings a pound (1795), a price attained also for some of the sixteen fish caught in 1798, and for 240 lb of salmon in 1803; 6 shillings a pound was paid for 20 lb and 33 lb fish (1802, 1806), 7 shillings and sixpence for a 23-pounder (1807), and 8 shillings for an 18 lb salmon (1808). Lower prices than these cited were also received; 2 shillings and sixpence a pound for 209 lb in 1801 (this was one day's catch), 2 shillings a pound in 1802, and in 1806 a 32 lb fish 'was spoiled by going to London by a slow coach; instead of fetching 6s., it fetched only 2s. per lb' (Venables, 1874). Prices varied with the condition of the fish, some of those caught in the bucks were dead when recovered and having been exposed for some time commanded a lower price than freshly killed fish. No doubt there was also variation with seasonal demand, and the quantity available elsewhere, but two general trends are noticeable, the steady upward progression of price per pound (a phenomenon of all food prices during this period), and the generally lower prices at Maidenhead compared with the tidal Thames (probably due to the latter's proximity to the London market). However, as the rent of the 'bucks' in which the salmon were caught at Taplow Mills was £24 a year, in most years the salmon alone brought in a handsome profit and other fish (such as lamperns and eels) provided additional income. The Lovegroves also cultivated extensive oyster beds in the vicinity which produced the material for their and other fish traps.

TABLE 3.1 *Summary of the Lovegroves' record of salmon caught at Boulter's Lock*

Year	*Salmon caught*	*Weight range*				*Total weight*		*Average weight*	
		lb	*oz*	*lb*	*oz*	*lb*	*oz*	*lb*	*oz*
1794	15	11	8	– 25	0	248	8	16	8
1795	19	?		– 21	0	168	0	8	13
1796	18	12	0	– 37	0	328	0	18	3
1797	37	?		– 31	0	670	8	18	2
1798	16	19	0	– 28	0	317	0	19	12
1799	36	?		– 28	0	507	0	14	1
1800	29	?		– 23	0	388	0	13	6
1801	66	6		– 37	0	1,124	0	17	0
1802	18	?		– 25	0	297	0	16	8
1803	20	?		– 31	0	374	0	18	11
1804	62	3	8	– 32	0	943	8	15	3
1805	7	?		– 21	0	116	0	16	8
1806	12	?		– 33	0	245	0	20	6
1807	16	?		– 30	0	253	0	15	12
1808	5	?		– 20	0	88	0	17	10
1809	8	?		– 32	0	116	0	14	8
1810	4	?		– 18	0	70	0	17	8
1811	16	4	8	– 16	0	181	12	11	6
1812	18	6	0	– 19	8	224	0	12	6
1813	14	6	0	– 22	0	220	0	15	11
1814	13	?		– 12	0	97	8	7	8
1815	4	?		– 14	0	52	0	13	0
1816	14	3	8	– 29	0	179	0	12	12
1817	5	4	12	– 29	0	76	4	15	4
1818	4	9	8	– 16	8	48	8	12	2
1819	5	10	0	– 23	0	84	0	16	12
1820	—	—	—	—	—	—	—	—	—
1821	2	13	0	– 18	0	31	0	15	8
Total	483					7,447	8	15	7

The journal also suggests that the figures in the decade before it was begun would have shown higher catches still. Thus 'about the year 1780 . . . my father caught upwards of 50 salmon in that reach opposite Cliveden Spring (about one and a half mile above Boulter's Lock), and the other fishermen caught in equal proportion, I remember . . . catching a salmon in the buck pool on 26th June, 1793, that weighed 42 lbs; length, four feet one inch' (Venables, 1874). The last salmon to be caught at Boulter's Lock was taken about the year 1823 or 1824; it was taken in the deep water above the lock.

From the details of months of capture given by Venables (1874) some idea of the season of capture is available. Before 1805, when the number caught began to fluctuate wildly, fish were caught in small numbers in April and May, the most productive months were June and July, August saw fewer caught, and the last few were taken in September. Many of the smallest fish were caught in the last two months.

What caused the catastrophic decline in the Thames salmon population? While a detailed account of the crisis in Thames fisheries is given elsewhere (p. 63), the fate of the salmon deserves special mention here. The Lovegroves' journal records the building of Windsor Lock in 1798, leaving open only 30 feet of the main river channel. The introduction of pound-locks by the canal builder and engineer James Brindley (1716–72) was a major contribution to the improvement of navigation in rivers and canals (see page 13), but one which had far-reaching consequences on the biology of rivers. To render the river deep enough for navigation throughout the year weirs were built across its width, interrupted by a pound-lock to allow vessels to pass through. From the late eighteenth century through to 1815 the Thames suffered the construction of a series of major weirs and adjacent locks. Teddington weir was built in 1811, Sunbury and Shepperton a year later, Chertsey in 1813, Penton Hook in 1814, and Hampton Court in 1815. Not only did the weirs represent a series of barriers across the river which could only be surmounted by leaping, although salmon can and do pass through locks as the gates are opened to allow vessels to pass, they also made the migrating fish more vulnerable to capture as they congregated in the weir pools. Their indirect effect was more serious. The structure of the river was altered; spawning grounds which require to be gravel shallows with a good flow of water were flooded with slow moving water caught behind the weir and quickly silted up. In some cases the gravel spawning beds were dredged out to improve navigation.

The erection of weirs and locks came at a time when the pollution of the middle reaches of the tidal Thames was serious. The combination was too severe a burden for the salmon and other migratory fishes, which during the first half of the nineteenth century passed from a moderate abundance to extinction or near extinction.

A further threat to the salmon was the activities of fishermen who, like the Lovegroves, operated a trapping system at a weir. The bucks in which the salmon were captured could be fished throughout the migrating period and operated to take advantage of the habit of the salmon returning to spawn in attempting to pass upstream wherever the downstream current allows. A small 'gate' above the buck was opened slightly to permit more water to flow and the salmon swam or leaped into the strengthened stream and were trapped. Where weirs were fitted with bucks across their width (as seems to have been the case at Taplow Mills) only a small proportion of migrating salmon could get upstream. As Richard Lovegrove himself admitted, 'there is no doubt that the bucks in the lock were very destructive, and there were not many fish got above them.'

The date of the capture of the last Thames salmon is difficult to give with certainty. By the 1820s they were very rare. At the beginning of this decade a 20 lb fish was caught 'in a deep hole near Surley Hall, just above Windsor. This was sold to the King for a guinea a pound' (Day, 1887) – the fisherman who caught it was (appropriately) named Finmore. Thirty shillings a pound were offered for Thames salmon for the coronation of King George IV in 1821 but none could be obtained – although a day later two were caught at Bugsby's Hole between Blackwall and Woolwich Reaches (Day, 1887). At Boulter's Lock, as we have seen, the last salmon was caught in 1823 or 1824. One was taken by a Mr Wilder at Monkey Island (between Taplow and Windsor) in 1830. William Yarrell (1836) reported a Thames salmon in June 1833 but gave no locality for its capture; if it was upstream of London it certainly qualifies as the last Thames salmon. Downstream, however, salmon were caught at intervals during the remainder of the nineteenth century. By this time, however, their capture had become newsworthy and witnesses tended to remember dates and weights and comment on them, perhaps not very accurately, years later. Thus the editor of the *Field* (27 April, 1867) commented that some six years prior to 1860 'we remember five fish in the period' at Erith (11 lb and 16 lb), Dartford Creek (2 fish), and another at Southend-on-Sea. Another report dated a capture at Erith as on 20 October 1860, while Francis Francis (a well-known Thames angler and fish conservationist) in the *Field* (18 August, 1866) reported the Erith catches as 10 lb and 16 lb and being made in November 1860, and the Dartford Creek fish were said to be grilse, one of which weighed 6½ lb. These probably all referred to the same fishes. Erith, on the Kentish side of the Thames, appeared to be a popular haunt for salmon, for on Thursday 25 October 1861 a Mr William Flynn saw a 20 lb salmon, caught at Erith, in the Pimlico shop of Mr Charles (fishmonger to Her Majesty) (Salmon Fisheries Commission, 1861). However, for reasons not specified, Mr Flynn doubted whether the fish had been caught where alleged, and clearly suspected it was a hoax. It might be suggested that Erith Reach represented the possible limit for salmon to penetrate up the Thames in the polluted state of river within the city; it lies 25·7 km (16 miles) below London Bridge.

A stuffed salmon was offered for sale in a London antique shop in October 1970 which, according to the painted inscription on the glass, was caught at Gravesend in May 1865 and weighed 23 lb. Frank Buckland had had it mounted but curiously never seems to have mentioned it in his varied writings, although the weight and locality, but not the date, correspond to a fish caught in 1870 referred to by Williams (1946). Murie (1903) recorded the capture of a salmon in a trawl off Southend in 1864, and others off Southend in July and in the Long Reach, below Purfleet, in August 1866 (the latter was 26 inches long and weighed 7½ lb). Another was 'seen about Sunbury Weir in June 1867' (Murie, 1901), but as the fish was not caught and identified positively there is a strong element of doubt about the record. Other captures were made in 1875 off Southend and Leigh, in May 1880 off Canvey Island (27½ lb), and in June 1891 at the entrance to Hadleigh Ray (33 lb), both these fish were sold to local dignitaries for the amazing price of one shilling a pound. A salmon caught on 30 October 1901 at Hadleigh Ray, near Canvey Island, which was first reported as a sea trout by Murie (1901) but later confirmed as a true salmon (*Field*, 1901), was 4 lb 7 oz in weight and 24 inches long and was in very poor condition. There seems little reason to doubt that in the outer Thames estuary salmon have been caught irregularly ever since these dates. Thus one of more than 20 lb was taken off Leigh in 1933, and in 1932 three were taken off Southend (January – two; April one of about 7 lb), while a 25 lb salmon was caught off Canvey in a whitebait net by the fishing bawley *Alice Matilda* in June 1937 (*Port of London Authority Monthly*, 1933–7). Clearly these outer estuary fish were not Thames salmon, but were migrants or stragglers from stocks in English Channel or North Sea rivers, and as such are likely to be captured in any bight in the North Sea – as they are today from time to time as, for example, in the Wash.

While smelt and salmon fisheries were of great value in the tidal and middle Thames reaches, the eel provided fisheries of at least equal importance along the whole of the Thames and its tributaries, although unfortunately few records of catches have been preserved. The eel is migratory but catadromous, breeding in the sea, the elvers entering river mouths after an almost transatlantic migration to grow for several years in freshwater. Quite possibly the fact that it did not spawn in the river (obviously gravid eels are very rarely seen), although this was not proven until the late years of the last century, was the reason that few conservation measures were enacted. There were, however, regulations against the use of the eel spear, and defining the number and season of eel-leaps or traps, but these were mainly to protect the young of other fishes.

Eels were captured mainly in the traditional eel-bucks, wicker baskets fixed to weirs, in mill-streams, or occasionally free standing (Yarrell,

FIGURE 3 *Eel-bucks*

1836), and in grig-weels. The eel-buck faced upstream to catch the eel as it migrated seawards (the opposite way to the salmon buck which caught fish ascending the river). Cornish (1902) described the capture of large eels in the grid of a mill at Dorchester on the River Thame, and any water-mill or weir might be expected to have had a similar capacity to catch eels. Eels and millers simply went together. He also described the 'grig-weel' sold by an old river hand at Eton, grig being the Thames-side name for young 'broad-nosed' eels (early naturalists, including William Yarrell, recognised several species of eel mainly on the basis of features due to changes during growth), which were baited wicker-work, narrow-necked traps for catching eels. The bucks were fished at night mainly during autumn, the main season on the river was November to January, although in the tributaries it started in October, when the maturing eels moved downstream on the first rush of water from heavy rain. Bucks were the main means of capture on the river and their use is hundreds, probably thousands, of years old. Wilson (1977) quoting from the Bailiffs accounts for the Manor of Cookham for 1287 referred to thirty shillings being obtained for eighty 'stikkes' of eels (a stikke contained twenty-six eels). This was a substantial sum of money at this period.

Mr Wilson further cites use of the bucks on the side channels of the main river (he writes of the river on the Buckinghamshire–Berkshire borders) until the twentieth century,

> There were usually six or seven bucks to a stage, with the open ends of the baskets, up to ten feet in diameter facing upstream. Side chambers were provided at the lower ends of the baskets, into which captured eels found their way out of the rush of the main river (Wilson, 1977).

According to Cornish (1902) 'often hundredweights are taken in a night, all of good size, one of the largest of which there is any record being one of 15 lbs, taken in the Kennet near Newbury.' It would have been of interest to have known of what date Cornish was writing. Eel traps were set in three different streams near Hampton Court around 1830 and were 'invariably supplied with eels' between November and January except when the frost was severe (Jesse, 1834). Bucks were fished at Taplow Mills for eels, the fishermen there reported that they weighed between ¾ and 2 or 2½ lbs, but in 1861 they had 'failed very much the last 30 or 40 years. There have not been so many taken between us and Reading.' However, as late as 1875, one stage of bucks could take above 56 lb of eels during a night (Wilson, 1977).

The reverse migration, when the elvers made their way up river, took place in spring (March to May, or June in the headwaters) in the Thames. Elvers are slender, about three inches in length and were present in huge numbers in the early nineteenth century and before. Edward Jesse (1834), then Surveyor of His Majesty's Parks, Palaces, etc., wrote of tracing the procession of elvers from near Blackfriars Bridge as far up river as Chertsey and presumed they made their way 'as far as or further than Oxford'. When the gates were shut at the locks at Teddington and Hampton Court they climbed the large posts of the flood gates, 'those which die, stick to the posts; others, which get a little higher, meet with the same fate, until at last a sufficient layer of them is formed to enable the rest to overcome the difficulty of the passage'. In 1832 Dr William Roots of Kingston kept special records of the eel-fare (as the elvers' migration was termed on the Thames). A few elvers had appeared at Twickenham early in April but the main run did not start until the afternoon of 30 April. For three days and nights they passed Kingston Bridge in considerable numbers. During one of the quieter periods Dr Roots stretched a thread of line over the margin of the river and counted the elvers crossing it, six hundred a minute, but an hour afterwards there were three times this many. Another estimate elsewhere on the river was of 1,600 a minute passing a fixed object. These estimates were made in daytime but the main migration takes place at night, so they are almost certainly minimum numbers. Even if the average over twenty-four hours is taken as 1,000 a minute, the total runs into nearly 1½ million per day passing this one fixed point on the bank of the Thames.

The eel-fare on the river was an occasion of some note. The Thames fishermen were said to keep a sort of holiday when the elvers first appeared in the river. The eel was a fish of considerable value which contributed greatly to their living, and the arrival of the young fish (which were no doubt caught and eaten in quantity – as they were on the Severn) was reasonably seen as an assurance of future income.

It was a living which was then in jeopardy, for as Dr Roots noted, 1832 was an unusually good year for elvers, better 'than I had witnessed for a long time before, indeed I may say since the building of the Lock at Teddington, or certainly since the year 1824'. As with the other migratory fishes the navigation locks had reduced the stock of eels, a trend which later pollution would accelerate.

Seawards of London there are very few reports of eels in the river, but judging from their general distribution they were probably found in numbers all along the river to the mouth of the estuary. As late as the 1920s they were caught on the River Roding as far up as Fyfield, near Ongar, on night-lines (Waters, 1964). Certainly, at the end of the nineteenth century Murie reported them as abundant in the Southend and Leigh-on-Sea areas, although there was little fishing effort directed towards catching them. At this time the middle tidal reaches were severely polluted and the Dutch eel schuyts which brought living eels from Holland in live wells or floating pierced boxes ('coffins') moored at Holehaven and sent boxed eels to the Billingsgate market. Frank Buckland (1883) writing around 1860 recounted a visit to a Dutch schuyt or eel ship (which he characteristically called a 'skoot'), during which the master told him that it was now many years since eels would live at London Bridge. 'The boats have gone first to Erith, then Greenhithe, and now they cannot come up further than Gravesend without killing the fish.' It is interesting that despite the wealth of eels in the Thames before it was badly polluted, the Dutch trade in eels was long established before 1412 when the Lord Mayor ruled that henceforth all eels bought from 'eleshippes' were to be sold by weight (Cameron, 1970). Yarrell (1836) commented that the London market was principally supplied from Holland, and that some ten vessels were engaged in the trade, each carrying some 15,000 to 20,000 lb per trip. It is said (Cameron, 1970) that nearly 10 million eels were sold each year at Billingsgate around 1861 and of these nearly three-quarters came from Holland. This confirms Yarrell's estimate of the importance of the Dutch fishery to the London market.

The lampern is another migratory fish which at an early date was the subject of a valuable fishery in the Thames. Like its relative, the sea lamprey, it spawns in freshwater having entered the river after a feeding period in the sea. The lampern and the lamprey are primitive fishes which lack a bony skelton. They are eel-like in shape, but have no gill cover, the gills opening to the outside through a series of holes along the anterior sides. They also lack jaws, although they have a rounded sucker disc which is equipped with teeth by means of which they bore through the skin of fishes and suck the blood and body fluids from their prey. The lampern spawns in spring on a gravelly spawning bed in which a shallow nest is excavated.

Although popular history credits the sea lamprey with being the food-fish (it is said that Henry I died as a result of a surfeit of lampreys, and the city of Gloucester annually presented a Christmas dish of lampreys to the monarch) in the Thames it was never abundant enough to be fished for commercially. Indeed it is doubtful whether this species was the one involved in either instance, the lampern being the more common, the more edible, and certainly not so repulsive looking as the larger, yellow-blotched sea lamprey.

Regulation of the season for catching lamperns was established in 1630 by the Lord Mayor of the City of London; 'no Fisherman . . . shall lay in the said River of *Thames* any Lampern-Leaps to take Lamperns before *Bartholomew-Tide* yearly, and so to continue till *Good Friday* . . .' (Binnell, 1758). Lamperns were, however, abundant at the time Thomas Pennant wrote (1791). He said they were taken in 'amazing quantities between Battersea Reach and Taplow Mills (a space of about fifty miles)'. Previously between 120,000 and 1 million were caught annually.

A fishery on this scale continued through to the middle 1800s. Thus in 1861, 60,000 were caught at Teddington (and sold for £3 a thousand). Most were caught at Teddington, Hampton Court and Molesey, where the weirs made them especially vulnerable, which suggests that the old methods of fishing 'lampern-leaps' had been largely abandoned by then. At Boulter's Lock, Taplow, there was a report of a single year's catch of 120,000 (Salmon Fisheries Commission, 1861), but after 1814 the catch decreased noticeably there, although some were recorded as sold in 1820. Here the lampern catch was made in the bucks at the weir, but once similar traps were used at Teddington the catch declined. As with the salmon, the lamperns decreased in numbers in proportion to the number of weirs (and fishing bucks) they had to encounter in the Thames. However, lamperns continued to run up the river and in 1882 were exceptionally abundant. The level of the weir at Teddington had been raised and the pool deepened and upstream access for the fish was more difficult, if not impossible. 1882 was the boom year at Teddington and fishermen there by mid-January had caught 120,000 lamperns, but upstream the fishery at Chertsey was ruined.

The value of the lampern fishery was considerable. It was estimated at £4,000 a year at Teddington alone in about 1878, and the price was £3 a thousand in 1861. When Pennant (1776) wrote he reported a price of £2 a thousand and a fishery of 450,000 annually, but in 1791 the price had risen to £3 a thousand (and from £5 to £8 a thousand for short-term contracts).

Curiously, as the lampern was reckoned to be a good food-fish, most of the Thames catch was purchased by Dutch fishermen for use as bait for the Dogger Bank cod and turbot fishery – both at that time caught entirely on long-lines. The English fishermen at Harwich had also been using lamperns as bait in the 1770s (Pennant, 1791), when they purchased 100,000 in one year.

Another abundant migratory fish which was captured in great numbers, although its value was not high, was the twaite shad. A relative of the herring, sprat, and pilchard, it bears a general resemblance to all these but is much larger, attaining a length of 20 inches, is deeper in the body, and usually has a line of dusky blotches along the side behind the head. It enters rivers from the sea to spawn upstream at about the extreme limit of tidal influence. It migrates upriver in May and in places is known as the May-fish; it spawns in June and July.

For many years there was considerable confusion as to the correct name for the two species of shad in British waters (a second, larger and now much rarer, species – the allis shad—is known), and whitebait, which formed an important fishery in the Thames, were held by some authors to be the young of the shad. Certainly young shad did occur in the whitebait. The earliest writers tended to confuse both species, but by the time William Yarrell wrote in 1836 there was no confusion. He wrote,

> Twaite shads appear during these three months (May–July) in abundance in the Thames, from the first point of land below Greenwich, opposite the Isle of Dogs, to the distance of a mile

> below; and great numbers are taken every season. . . . Formerly great quantities . . . were caught with nets . . . opposite the present Penitentiary at Millbank, Westminster. Above Putney Bridge was another favourite spot for them. . . .

Unfortunately, despite their abundance they were then of little value to the fisherman, as Yarrell put it, 'being in little repute as food, their muscles being exceeding full of bones and dry'. It is probably significant, however, that shad remains have been found in several medieval archaeological sites in London. Moreover, as numerous regulations existed in the seventeenth and eighteenth centuries regulating the fishery of 'shadds, in shadding time', it clearly had been an important food-fish, even if it did not appeal to the same palates as the salmon.

By the end of the nineteenth century shad were not found in the middle reaches of the tidal Thames or upstream. Even as far down as Gravesend in 1878 a witness with twenty years' experience as a fisherman claimed that there were 'few shad among the whitebait. They have fallen off wonderfully' (Salmon Fisheries Commission, 1861). Murie (1903), writing of the Thames estuary, claimed that they were still abundant, and cited a number of captures at Queenborough, the Blyth Sands, Leigh, Hadleigh, and Southend. It is doubtful whether these fish spawned anywhere in the Thames estuary (although they may have done so in the Crouch and Blackwater to the north and in the Medway, although there is no proof of this). Probably these fish were stragglers from spawning stocks in other rivers, as are the shad which are regularly but sparsely captured around Southend and in the lower estuary today.

FRESHWATER FISH AND FISHERIES

In view of the numbers of fishermen who earned their living on the Thames and have left observations on their catches of the prime migratory species, it is obvious that many freshwater fish must have been caught. That they have by and large gone unrecorded is probably due to their low marketable value which would have meant that, like shad, they were sold locally (not marketed through Billingsgate) or were kept for the consumption of the fisherman and his family.

That freshwater fishes were eaten widely in the seventeenth and eighteenth centuries, before modern transport methods brought sea fish fresh to inland areas, is indisputable, as the remains principally of roach have been found in several medieval archaeological sites. The Corporation of London, exercising its control on the London free-fishing made numerous regulations concerning close seasons and fishing methods, some quite specifically aimed at protecting what are now known as coarse fish. For example, 'no Fisherman . . . shall at any time of the Year use or exercise any Flue, Trammel, double-walled Net, or whatsoever, for that they are not only the utter Destruction of all breeding Barbels . . .', and 'no Fishermen . . . shall lay any Weels called *Kills*, in any Place of the River, from the 10th of *March*, till the 10th of *May* yearly, for that all Roaches do then shed their Spawn' (Binnell, 1758). The same regulation prohibited the cutting of bullrushes, flags or sedges in the river between Richmond and Staines Bridge 'for they are a great Succour and Safeguard unto the Fish'; an extraordinarily perspicacious regulation for the time (1630).

The Thames system contains a rich freshwater fish fauna which in the natural state would be prevented from moving downstream only by the incursion of salt water with tidal movement. While the old, many-piered London Bridge stood (until 1832), the effects of the tidal flow virtually ceased there, upstream being freshwater and downstream freshwater (at low-tide conditions) to slightly saline at high tide with increasing salinity further downstream. The 'barrier' effect of old London Bridge made it possible for the river to freeze over in severe winters, and was responsible for the danger in 'shooting the bridge' by water at low tide. Nervous passengers passing downstream by boat left it at the steps upstream of the Bridge and walked down to join the waterman at the next lot of steps. Its effect on the fishes was marked, for those like the barbel that could not tolerate salt water were not found below the bridge. Upstream of the bridge, barbel, roach, dace, and 'bleak in great plenty' (Pennant, 1791) were found; downstream relatively few freshwater fishes were encountered.

Freshwater fishes in the Thames enter the literary record only in the mid-nineteenth century. Possibly this was because, with the destruction of the most valuable fisheries for salmon, smelt, and lampern, fishermen upstream could only catch the non-migratory species, but also because with social changes the fishing began to attract a new, literate class who angled for fish as a leisure activity. Thus in 1838 the Thames Angling Preservation Society was founded by several anglers of relatively high social standing (Cargill, 1972). A year later the Society was empowered by the City of London Corporation to appoint five extra bailiffs, in addition to the single city-appointed bailiff to enforce the Lord Mayor's 'Rules, Orders and Ordinances' for fishermen upstream to Staines. At this time netting was carried on by both professional fishermen and by 'poachers' (presumably fishermen who had not served an apprenticeship on the river and were thus unlicensed), but neither class always observed the regulations for close seasons and mesh size, or respected the preserves where fishing was not allowed. The Thames Angling Preservation Society's bailiffs soon made their presence felt; between 1839 and 1844 they secured eighty-four convictions, fines amounting to £138, and twenty-nine illegal nets were burnt. Several illegal netsmen were imprisoned. Bailiffing in these times was a dangerous occupation and all bailiffs were armed with truncheons!

The Thames Angling Preservation Society eventually secured the cessation of netting between Richmond and Staines, the abolition being incorporated in the Thames Conservancy's bye-laws in 1860, and later (1886) were instrumental in extending the prohibition downstream to Isleworth Church. It remained legal to net the river, subject to season and to the net-satisfying requirements of mesh size and length, below Isleworth to the mouth of the river. Before netting was made illegal, the Society was permitted by the City Corporation to set up a preserve extending to 1,793 metres (1,960 yards) between Broom Hall, Teddington, through the backwaters to the railway bridge at Kingston. Netting in the preserve was made impossible by sinking in it five punts, two iron waggons, two punt loads of old iron gas lamps 'and other useful things', as well as driving 450 stakes into the bed (Brougham, 1893).

The Society was also instrumental in establishing a fish hatchery on the Thames, a venture in which Frank Buckland and Mr Stephen Ponder joined and later prosecuted on their own. Buckland and Ponder built a 'hatching apparatus' in Ponder's greenhouse at Hampton in 1863 (Buckland, 1863) and persuaded the Thames Conservancy to put aside the stream running parallel to the lock at Sunbury as a rearing place. Into this stream in February 1865 some 30,000 young trout were released (after a champagne lunch at the Red Lion Hotel, Hampton, for the press and the parties involved). The Society itself had installed fish-hatching boxes in the Christian Spring at Hampton, but eventually abandoned the project. Buckland and Ponder's more elaborate arrangements were continued for some years, and other enthusiasts (Mr James Forbes at Chertsey Bridge and Mr Thomas Spreckley at Sunbury) later set up hatcheries for trout, and even perch, rearing. Francis Francis was closely associated with both the Society and Stephen Ponder's experiments, and seems to have been an organiser of both ventures.

Buckland (1863) reported that Mr Ponder's apparatus had been instrumental in hatching eggs from which 35,000 fish (salmon, trout, charr and grayling) were released into the river. Murie (1903) gave a summary of the Thames Angling Preservation Society's hatchery in the five years between 1861 and 1866, when some 48,000 salmon, 4,000 sea trout and 137,000 trout were released to the Thames. Unless some of the few salmon which occurred in the Thames in the 1860s were the survivors of these stocks (and there is no reason to suppose they were) all 48,000 must have disappeared without trace.

A later attempt to reintroduce salmon to the Thames was made by The Thames Salmon Association between 1901 and 1906 and 100,000 young salmon, mostly parr, but 700 smolts in 1901, were stocked into the Thames at Teddington, again without success (Williams, 1946). This same body was responsible for the introduction of the huchen (*Hucho hucho*), a native of the river Danube, and a relative of the salmon and trout. In 1906 a large number of two-year-old huchen (hatched from 20,000 eggs introduced in 1904) were liberated at Nuneham by the Abingdon Restocking Association. The fate of these fish is not certainly established, but in the 1920s and 1930s there were numerous reports of huchen of 30 to 40 lb being seen or hooked by anglers, although not landed, and these rumours were sanctified by the chief inspector of the Thames Conservancy in his report for 1933, who claimed that there were huchen in the Thames up to 40 lb in weight (Pearson, 1961). The truth is that although there was intense publicity, numerous anglers trying to catch the elusive huchen, and even a special netting of the Thames arranged, no huchen was ever landed and positively identified. The only evidence for their survival were eye-witness accounts of large fish seen in the water and anglers who hooked a large fish and lost it when their tackle broke, and this is not sufficient proof that the huchen survived its introduction to the Thames.

Trout were a popular angling quarry, partly because of their size, and figured largely in the stocking of the river. Teddington Lock yielded several weighing 14–15 lb between 1854 and 1857 (Wright, 1971). Henry Farnell in 1861 reported catching Thames trout of 10 and 11 lb, and knew of one at Bell's Weir (just above Staines) of 15 lb, and John Gould giving evidence at the same enquiry said, 'I have killed many scores (of trout) between 9 and 10 lbs . . . the largest Thames trout I ever saw was between 16 and 17 lbs' (Salmon Fisheries Commission, 1861). Possibly trout were less common in the main tidal river than in the tributaries and upstream. Thus Richard Lovegrove of Taplow Mills spoke of 'an immense lot of trout there' some weighing up to 13 lb, and John Gould recalled that up to 1850 the Wycombe stream was filled with magnificent trout, so that 16 to 18 fish from 1½ to 3 lb could be caught in a day, and on the River Chess in 1861 trout were numerous and he had seen 100 brace of fine trout at the tail of a mill-pool. Sadly, the remarkable trout of the River Wye in Bucks were killed off by pollution from the paper-mills in the 1850s (Wilson, 1977). The upper Lee was also renowned for the fine trout it contained.

Other freshwater fishes which were captured by anglers and netsmen included barbel which, being large (it grows to a weight of 13 lb) was netted with some ease, pike, perch, chub, roach, and dace. The catches were often large. Around 1855 60 lb of perch were caught by angling in an hour at Putney Bridge (Wright, 1971). At the beginning of Francis Francis's angling career, a summer's day spent fishing at Richmond produced 100 lb of fish, including 16 barbel up to 4 lb, and several hundred roach and dace, while at Molesey weir he caught 22 pike in two days.

ESTUARINE FISH AND FISHERIES

Downstream of London the Thames was fished for a number of species of fishes and some invertebrates. Changes in the regions fished resulted from the poor quality of the water during the nineteenth century, so that eventually the main estuarine fisheries were prosecuted only in the mouth of the river. These changes must have resulted in considerable social alterations amongst the Thames-side fishing communities which were not well recorded in the literature, although some information is available for the Barking, Leigh-on-Sea and Gravesend fisheries. Possibly one reason for this sparsity of information is because the changes in the river and its fisheries coincided with major developments in transporting fresh fish to the London market – so that the population of London was in general unaffected. Had the fisheries failed a century earlier there would have been severe economic repercussions – which might have left fuller records.

Several important estuarine fisheries need to be discussed, chief amongst them that for whitebait – the Thames fish and London dish *par excellence*. The early history of the fishery is obscure and the true identity of whitebait was for long uncertain. Thomas Pennant (1776) regarded the whitebait as the young of the bleak which was, as it is now, a very abundant freshwater fish in the Thames, but he conceded that others considered it to be young shad, sprat, or smelt. Edward Donovan, in his

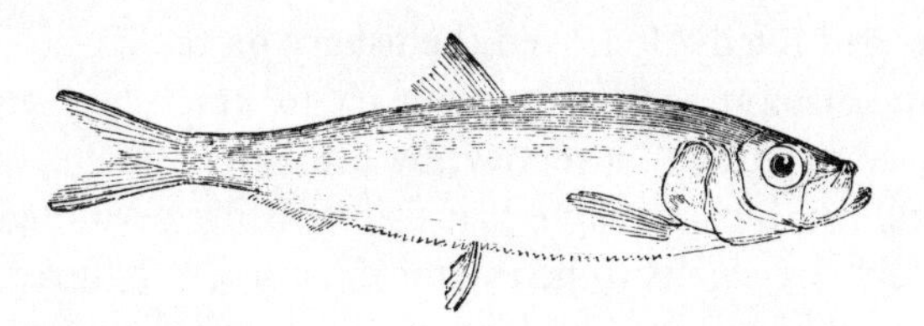

FIGURE 4 *Whitebait*

Natural History of British Fishes (1802–1808), declared that they were young shads. Both, of course, may well have been unfortunate in selecting a few fish from a sample of whitebait, for young bleak and shads could have been found in it. Confusion became more profound when Yarrell (1828) declared that the whitebait was a perfectly distinct miniature species of fish, a view that he further propounded in his *A History of British Fishes* (1836) where he referred to it as *Clupea alba*. Yarrell's arguments were evidently so persuasive that the distinguished French naturalist Achille Valenciennes (1847) decided that it was sufficiently distinct to be placed in a new genus, *Rogenia*. For a short while *Rogenia alba* became the scientific name for the Thames whitebait.

A later generation of naturalists, however, reviewed the whole matter and came to the conclusion that whitebait were the young of sprats and herring varying in proportion with place and season, but often with a variety of other small fishes in addition. Sprats dominated in the catch from February to April, herrings in June and July, while through May both species were equally abundant.

Possibly Yarrell was influenced by the opinion current among London fishermen that the whitebait was a distinct and fully grown fish. But it was in the fishermen's own interest to have the weight of zoological authority on their side, as it freed them from the charge of destroying young fish. In fact, the main prohibitions regulating the whitebait fishery seemed to be aimed at preventing the wholesale destruction of young fish in the river, viz. 'Whitebait shall only be taken with a *Wade-net* not exceeding four Yards in Length, and with Meshes of not less size than three Quarters of an Inch, from Knot to Knot . . .' (1757), and '. . . it appearing to this Court that under pretence of taking *White-bait* the small fry of various species of fish are destroyed' (Murie, MS). This, indeed, was the allegation of Mr Sanders, the City of London's water bailiff and a deputation from the Body of Fishermen on the River Thames who presented a petition to the Lord Mayor about 19 April 1828.

The petition set forth that the taking of these little fish, called white bait, had totally destroyed the fishery on the River. . . . At the present season it had been customary for the fishermen to earn a considerable sum of money, by fishing for smelts, which were lawful fish, but the white bait fishing being permitted, there were scarcely any smelts to be caught . . . ; it also reduced the supply of chad [shad], roach, dace, and plaice [? flounder], and caused a scarcity in the market, and thus increased the price of them.

The Lord Mayor said he was fully aware of the injurious tendency of this white bait fishing, which was carried on for no other purpose than to pamper the dainty appetites of a few at the expense of the ruin of hundreds of honest and industrious men. He was, therefore, determined to exercise all his power to put a stop to it.

It is certain that around this period the whitebait was the subject of important fisheries and suffered from a fashionable popularity. Thomas Pennant (1776) wrote,

During the month of July there appear in the *Thames*, near *Blackwall* and *Greenwich*, innumerable multitudes of small fish, which are known to the *Londoners* by the name of *White Bait*. They are esteemed very delicious when fried with fine flour, and occasion, during the season, a vast resort of the lower order of epicures to the taverns contiguous to the places they are taken at.

His rather disdainful reference to the lower order of epicures is difficult to explain unless, perhaps Pennant, a country gentleman from Flintshire, found the Blackwall region distasteful for some reason! Peter Lord of the Artichoke Tavern at Blackwall claimed to have done much to bring the whitebait into repute as a delicacy, and in 1765 was advertising 'his duty to acquaint the public that the much admired fish called white bait, are now just come in and are in great perfection'. A large whitebait dish with a vignette of the Artichoke Tavern from around this period is still preserved in Poplar Public Library.

The Artichoke Tavern flourished from about 1750 to 1870, and even at this later date visitors to the Blackwall and Greenwich regions were regaled with whitebait (although they were then presumably not caught in the river there) (Timbs, 1886). Other taverns in the Greenwich area (the Ship, and the Crown and Sceptre), and at Blackwall (Mr Lovegrove's West India Dock Tavern) specialised in whitebait dishes. The care with which the fish were prepared and cooked must have contributed greatly to its appeal to the epicure; they were kept in water, then completely enveloped in flour, shaken in a colander and 'thrown into hot lard contained in a copper cauldron or stew-pan placed over a charcoal fire; in about two minutes they are removed by a tin skimmer . . . and served up instantly, by placing them on a fish-drainer in a dish' (Timbs, 1886). The freshness of the fish when captured in the adjacent river no doubt also helped make the flavour better. The social standing of whitebait improved sometime after 1812 when the dining club of government ministers and other gentry that had had its origin at Breach House, Dagenham, moved upstream to Greenwich, because after the move whitebait was served at these dinners (Powell, 1966).

In 1836 William Yarrell remarked on the social progress of this dish between his and Pennant's time,

> at present, the fashion of enjoying the excellent course of fish as served up either at Greenwich or Blackwall is sanctioned by the highest authorities from the court at St James's Palace in the West, to the Lord Mayor and his court in the East, including the Cabinet Ministers and the philosophers of the Royal Society.

He elaborated his reference to the Cabinet ministers by quoting the *Morning Post* of 10 September 1835, which reported the visit of the cabinet 'to Lovegrove's West India Dock Tavern, Blackwall, to partake of their annual fish dinner'. By having their dinner so late in the year, the cabinet must have been served with small sprats rather than whitebait!

That whitebait was a speciality of the Blackwall and Greenwich regions suggests that the young fish were in abundance here. In fact, being nektonic (actively swimming near the surface) they would be carried upstream by the tide (returning downstream on the ebb) and would thus become concentrated by the sudden narrowing of the tidal stream as the river curves in Bugsby's Reach. There are, therefore, physical reasons why the fishery became established there. Although whitebait would have been present in season all down the river, they would not have been so concentrated and greater fishing effort was required to catch them. The fishing was to some extent governed by the tides: at Woolwich they were caught three or four hours after the tide had been running (Yarrell, 1836); further upstream they would have been captured for only an hour or so either side of flood tide.

The ebb and flow of the tide were utilised in catching the fish. The upstream whitebait net was fished from a boat moored in the tideway and the net was lowered over the side. The net had a relatively small mouth, only about three feet square, held open by two horizontal beams joined centrally by a third, vertical, beam. Despite its small aperture the net was very long and the mesh, especially towards the bag end, very fine. As the boat rode at its mooring so the net faced the flowing water and any fish entering its mouth continued with the tide into the belly of the net which flowed loose beside the boat. From time to time the hose of the net was shaken to drive any meshed fish further down, and as the catch accumulated the end of the net was taken into the boat, the cod-end untied and the whitebait shaken out. The net was fished mainly between four and six feet from the surface and was a facile tool in skilled hands, for whitebait are very sensitive to changes in depth and the level of the net could be varied until it was fishing in the densest concentration of fish. Yarrell's (1836) account of this fishing method, which is the main source of this description, was accompanied by an illustration, but it

FIGURE 5 *Whitebait net*

should be noted that the net was shown upside down (Buckland and Walpole, 1879). The traditional whitebait net was the forerunner of the much larger stow-net which was used downstream to catch both sprats and whitebait. By the 1900s the stow-nets used from the Southend bawleys for whitebait had a mouth 22 feet wide and 30 feet deep, while the hose was up to 60 feet in length. Fished for two to three hours in the early morning it could catch 3 hundredweight at a time.

In the late nineteenth century a drag-net (similar to a shore seine) was in use for whitebaiting in the lower Thames, and this was probably also used further upstream at an earlier period. This was similar in construction and use to the traditional peter-net. Murie's (MS) description of its use and figures are given here.

> According to circumstances such as depth of water, state of tide, etc., so shorter or longer haul-ropes are used. When, say, it is high water and close to a steep sea wall no extra endropes are bent on – the net itself being hauled in. At low tide and in shallow water though, lines from 15 up to 80 fathoms are bent on, so as to allow the net to go clear into the deeper water prior to hauling in.
>
> All being ready a man or lad is taken ashore and holds on to one end of the rope. Then the boat is rowed seaward at a right angle to the shore, while paying out the net as it proceeds. When the net is well out . . . the boat gradually swerves and makes a horse-shoe curve for inshore, nearing which the rope is suddenly slacked out and full swing given to oars for the men to get ashore quickly. One hand jumps out and pulls the net ashore, meanwhile the fisher at the opposite end of the net also pulls lustily. The third hand anchors the boat in position and thereafter . . . jumps into the water and by splashing drives the fish in the pocket.

Murie's account was based on a trip made with a Leigh-on-Sea whitebait boat one summer's morning (about 1903), starting at 3 a.m., during which ten hauls were made before 8.45 a.m. and the catch rowed back in time for the 9.40 a.m. train to London.

By the 1870s whitebait fishing had moved downstream to the vicinity of Gravesend, Southend-on-Sea and the Lower Hope. To some extent this was certainly due to pollution of the upstream reaches, but technical changes had also had an effect. The demand for whitebait had increased and the larger stow-nets were not suitable for fishing in the confines of a narrowing river in increased numbers. Additionally, railway transport meant that it was possible to catch whitebait, say, off Leigh-on-Sea starting at 3 a.m. and by means of the early train have it sold at Billingsgate that morning – exactly the treatment necessary for this fragile and perishable fish.

At the enquiry held by Buckland and Walpole (1878) for their report on the sea fisheries of England and Wales, numerous fishermen gave evidence of the harmful effects of whitebait fishing on the fry of other fish, and it is certain that a great deal of young fish of species other than herring and sprat were captured when the stow-nets were not properly handled. If fished too close to the river bed, for example, they caught many young flatfishes. Also this heavy exploitation of the young clupeids had resulted in such a drastic falling off of sprat catches that in order to make a living more fishermen were forced to fish for whitebait. At the time of the enquiry, one Gravesend man claimed that only three local fishermen were catching whitebait but about forty Greenwich men were fishing there; another man claimed there were sixty men whitebaiting at Greenwich.

The fishery in the Thames was very extensive by 1878. At Queenborough, Isle of Sheppey, forty baskets of whitebait were sent to London in one morning, each weighing between 40 and 56 lb, and a Queenborough fisherman had caught 2 hundredweight in four hours in the morning on which he gave evidence. At this time there were thirty to forty people employed in Queenborough alone in the whitebaiting industry, and £1,000 a year was coming into the port as fishermen's wages. At Southend there were forty fishermen fishing for whitebait which was the main fishing there between February and August.

The consequences of this fishery, essentially for young sprat and herring, can be evaluated from figures for the composition of a large whitebait box containing about 6 gallons which held over 16,000 young sprats and herring. During the season the proportions between the species varied. At the outset young sprats were dominant but towards midsummer were replaced by the relatively inferior 'yawlings' or herrings. At all times there was a generous admixture of other, particularly pelagic, species. About May especially the transparent goby known locally as 'Roshians' became common, so-called by the fishermen of the estuary because at the time of the Crimean war they were exceptionally abundant in the river (Murie, 1903).

The causes for the vast expansion in the fishery and its migration downstream have already been discussed. It was certainly a late development at Leigh-on-Sea and a Mr James Henry Cannon at Southend claimed that his grandfather, Richard Cannon of Blackwall, had 'invented' whitebait in 1780 – he also claimed the name whitebait was used because these fish were initially used to bait eel pots. Neither claim seems to be true! Certainly during the mid-1880s there was some relaxation of the City of London's bailiffing – for stow-netting upstream was illegal, and not until 1893 was legislation enforced to make fishermen work seaward of the Crowstone-Yantlet line. The failure to enforce the bye-laws prohibiting the taking of young fish, stemming in part from the insistence of fishermen and some naturalists that whitebait were adult fish of a separate species, permitted the huge increase in this fishery.

The sprat, a small relative of the herring, is especially abundant in estuaries and in inshore waters, and large fisheries have existed to exploit it. As, like the whitebait, it spoils quickly, the biggest fisheries have always been near large cities where the catch can be marketed soon after capture. In the Thames, the sprat fishery was of great importance in the lower reaches of the river, but apart from the whitebait, a fishery did not exist upstream. The adult fish enter the estuary during the early winter (traditionally the fishery off Southend started in mid-November, but if possible catches were made in time for Lord Mayor's Day – 9 November). Their movement upriver continued through December to March, but the extent to which they penetrated upstream depended on tidal movement and the weather. Following high spring tides, especially when backed up by north-east winds sprats could be found well upstream, but probably in general they were not particularly abundant above Long Reach and St Clement's Reach (near the present Dartford Tunnel).

The sprat fishery is of great antiquity. Emmison (1976) recorded an incidental mention in the will of a Leigh man dated 1558 of 'a firkin of sprats'. They were fished for off Gravesend in the mid-eighteenth century, and Pennant (1776) refers to their being cured there like red-herrings in 1776. The Thames-mouth fishery seems always to have been prosecuted using stow-nets, the bawleys working wherever the fish were abundant (often deduced by flocks of feeding birds and porpoises). This gear requires the fishing boat to be anchored while the fish swim into the open mouth of the suspended net. Huge quantities can be caught in this way. Murie (1903) records some astonishing catches; 300 bushels in a single haul, and another between the Knock Buoy and Southend Pier when 180 bushels were taken. But there were many nights when the stow-nets were fished with scarcely any catch to speak of, and in the 1898–9 season the working expenses of the boats were not covered by the income from the catch. This was probably due to a poor spawning season two or three years before, for the sprat is particularly prone to variation in abundance. It might also have been due to the heavy fishing for whitebait that had been carried on for several years in this area (an over-fished stock frequently fluctuates greatly in abundance). The evidence before the Commissioners for Sea Fisheries in 1878 was contradictory, a Southend man claiming that taking five-year averages there had been no decline for forty years, while at Gravesend sprats were said to have fallen off 75 per cent, and at Leigh they had decreased. Murie (1903) concluded at the close of the nineteenth century, that 'there is no gainsaying the fact, that as a paying occupation sprat fishing . . . is on the decline'.

The fluctuations in the catches, particularly the immense gluts that occurred, meant that the market price was often unstable. Once the stow-boat fishery became established in the 1850s, it was usual for the fastest bawley in a partnership to take the catch and make the run to Billingsgate, although if winds were contrary the catch might be marketed in Chatham or other large towns. Only as a last resort was the railway used, for that increased the costs unacceptably. In glut years the price varied (although in 1877 – a good year – one Queenborough man took £50 worth in a week) and much of the catch went to farmers for manure. The so-called muck boats between Gravesend and Erith on the Kent coast came out specially to load up with sprats for manure, and they fetched 7 to 11½ pence a bushel (as opposed to the Billingsgate price of 1 to 8 shillings a bushel – pre-1878 prices). This was not a new development, for Yarrell (1836) had commented on their use as manure by Dartford farmers. Although this use of sprats was much criticised, it arose largely as a result of the unevenness of the catches and it meant that the men received something for their work; the alternatives were either to stop fishing or to dump the catch, neither of which was as practical then as they would be today in similar circumstances.

Another fishery existed specially to provide manure for the farmer, the curious one for five-fingers (as they are locally known) or the starfish, *Asterias rubens*. Up to about 1850 this was a minor occupation for the Leigh men, about fourteen or fifteen boats being occupied in it in spring and autumn. Most fishing was done in the mouth of the estuary below the Nore and some of the most prolific localities were off the Isle of Sheppey, but on occasions five-finger dredging was tried opposite Shellhaven or near the Yantlet. Possibly the fishing had started as a means of controlling these serious predators of oysters and the catch being found to be marketable as manure, it became a minor industry in its own right. Each boat would get a cargo once or twice a week each of 5 or 6 tons, worth on average £1 per ton. Much of the cargo was taken to farmers by contract in the Leigh area but some went to Gravesend, Erith, and Mucking. The quantities concerned were considerable, one farmer on the River Blackwater recording the spreading of 70 tons, purchased at 16 shillings a ton, on his fields (Murie, MS).

The five-finger fishery continued in a small way through to the 1880s, but by then commercial fertilisers (guano) were widely available even to rural farmers, farming was locally depressed by a series of very wet years, and above all the whitebait, sprat and shrimp fisheries provided a better income. Occasionally fishermen still dredged during a season, possibly to supply some individual farmer's needs, which is not surprising in that two of the most conservative industries, fishing and farming, were involved. In the Thames it is doubtful whether starfish dredging continued into the 1900s.

In 1758 Robert Binnell wrote, 'there is no River in all *Europe*, that is a better, or a more speedy Breeder, and Nourisher of its Fish (particularly the *Flounder*) than is the Thames', and although he, as Bailiff to the City, may be suspected of special pleading, it does seem that the flounder was exceptionally abundant in the river and was fished for extensively. The flounder is the only flatfish in northern Europe to enter freshwater, and although it breeds in the sea the young fish are found in freshwater at small size and in great numbers. Large flounders can also be found in rivers, although they move downstream into the estuary as they become sexually mature.

Numerous bye-laws existed on the statute books of the City of London regulating the size limits, season and method of capture of flounders, which suggests that it was one of the more important fisheries. Due to its habit of entering freshwater it is likely that these fisheries were carried on along the length of the tidal river. Yarrell (1836) reported that it was taken as high upriver as Teddington and Sunbury, as well as in the mouth of the Mole where it joins the Thames at Hampton Court. He added, 'this species is caught in considerable quantities from Deptford to Richmond by Thames fishermen, who, with the assistance of an apprentice, use a net of a particular sort, called a tuck-net or tuck sean.' This tuck-net was in effect a seine or peter-net of greater than usual depth which was anchored at one end while in use. Peter-nets were also used regularly to capture flounders, and in the mouth of the river, plaice and dabs. Where the structure of the river bed allowed and certainly in the estuary where 'guts' or tidal inlets are numerous, stop-nets were commonly employed. Although they were illegal in most of the river and were classed as 'fixed engines' by the Act of 1861, a wall of netting was so speedily placed in position across the mouth of the gut at high tide, trapping all fish, but most effectively flounders, as the tide receded, that they were difficult to detect in out-of-the-way regions. Murie (MS) recalls that in the late 1800s a fisherman still

> openly credited with having regular recourse to Stopnetting in the guts and creeklets between Holehaven and the Crowstone as well as inlets on the opposite Kent side of the water, was a hoary old fisherman from Gravesend. He . . . annually paid visits to the said neighbourhoods during the summer and autumn months in his

antiquated 'Pink-Stern' craft, ostensibly for ordinary fish trawling. But to Hoskins all leniency was shown, and he stopnetted more in the runlets on shore than trawled in the deeper waters.

Flounders, and other flatfishes, were the mainstay of the catch of the 'kiddles' (also spelled kiddels, or kettles) and fish weirs. These permanent erections on the tidal shore were prohibited time and again in City of London regulations (and indeed in Magna Carta), so often, in fact, that it seems that the Bailiff of the City must have had great difficulty in bringing offenders before the courts. In addition to the alleged destruction of young fish, which was one reason for the City's concern, they were serious hazards to navigation in the river, for originally they were built of stout timber interwoven with osiers and brushwood. Their danger to shipping was the major reason for the legislation against them. In the estuary their design varied with the features of the shore, but essentially they represented a wide V-shape, the point towards the sea or deep water, equipped with a holding pound or box. The fish swam along and over the shore during high tide but on their return encountered the arms of the kiddle and following them down to deeper water were eventually trapped in the pound. Not just flounders, but a whole range of bottom-feeding fish which explore the intertidal zone during high tide, as well as migratory salmon and sea trout weie captured in these kiddles. In later days the brushwood was often replaced by netting.

With the increasing canalisation of the upstream Thames, the building of sea walls and decrease in natural foreshore and, not least, their illegality, kiddles were not employed within the City of London area. Downstream in the outer estuary kiddles were in use at Maplin, and on the Kent coast at Seasalter, in the East Swale mouth, near Whitstable, until the end of the last century (Buckland and Walpole, 1879). Murie (MS) quoting Benton's *History of the Rochford Hundred*, notes that the manorial rights of Foulness included the use of twenty-seven kiddles, each of which was known by its special name, e.g. Spalnet, Plecke, etc. The important role that the kiddles played in the lives of the inhabitants of Foulness and Wakering is illustrated by the care with which each was described and named in the wills of the islanders (Emmison, 1976). Names such as Saturday, Half Ebb, Pleck, Barnfleet, Crouch, South, Spedwell, and le Tepe kiddles occur in wills between 1580 and 1596; many of them recurring in the will of a legatee, John Stapell the elder, who left ten kiddles in 1604 to his son William. There are also references to 'a certain place for laying of hooks and taking of fishes' in the will of

John Staples 'the eldest' of 1586, which show that the foreshore was fished with set lines as well as the traditional osier kiddles (keddles seems to have been the normal spelling in the sixteenth century). Foreshore fishing and fowling, with a little grazing seem to have been the way of life for the islanders at this period, judging from the scholarly researches of Dr Emmison.

Another fishery which was of great importance to the economy of the seaward reaches of the river was that for the brown shrimp, *Crangon crangon*, also known as the common shrimp or grey shrimp to distinguish it from the pink shrimp, *Pandalus montagui*, which is common along the Kent and Essex coasts and occurs in the estuary. The brown shrimp is found on sandy and muddy shores, and tolerates estuarine conditions where the salinity is not too low. Two fishing communities on the lower Thames became associated with the shrimp fishery, Gravesend and Leigh. The origin of shrimp fishing is unclear, for although there was a bye-law in 1697 of the Thames Free Fishermen (modified by the City Corporation in 1757 and 1785) limiting 'dragging' for shrimps except between 1 November and Good Friday following, from 'Magget-Nasse' (possibly Margaret Ness between Gallions and Barking Reaches) seawards, there is little evidence as to how far upstream or how heavily shrimping was prosecuted. Certainly, this bye-law suggests that shrimping was possible above Barking Reach (if that is the identity of Magget-Nasse), but the development of Gravesend as the major shrimping port suggests that later the fishery was prosecuted downstream, possibly in the Gravesend and the Lower Hope Reaches. Gravesend was the prime shrimping port on the Estuary and still retained its importance into the 1920s when, however, the bawleys were fishing down towards Southend. So closely was shrimping associated with the town that the important Kentish brewers, Russells Gravesend Brewery Co, marketed a special beer known as Shrimp Brand of which the local newspaper, the Gravesend and Northfleet Standard of 12 February 1909, wrote that although 'many different brands of beer come and go no make has ever caught on like Shrimp Brand'. The Company's official seal incorporated an excellent likeness of the brown shrimp with the words 'Shrimp Brand' on either side. Although Russells Brewery was taken over by Truman, Hanbury, Buxton and Co. several years ago, memories of Shrimp Brand beer still linger on in the Gravesend area. Leigh men seem not to have taken to shrimping until about 1800 (Murie, MS) although again the evidence is not clear and the fishing grounds were almost certainly local. However, as late as 1878 stow-boats were fishing for shrimps off Greenhithe and

Northfleet Hope upstream of Gravesend (Buckland and Walpole, 1879). By the late 1800s Murie records shrimp trawlers 'constantly at work' on the Sea Reach fishing grounds, roughly from Mucking Light to the Leigh grounds, although then, and later, the shrimpers worked to seawards and even fished on the Harwich grounds – where the catch was chiefly pink shrimps.

Most of the early shrimping was done using a trim-tram or trunck-net which was characteristic of the middle Thames fishing. Descriptions of it imply that it was clumsy to use and most unhandy to have across the deck when not in use, with its 18-foot-long elm beam and the two diagonally forward-pointing, iron shod snooks to which the warp was attached. The net was fastened to the beam and its mouth held open by a socketed vertical stick, 5 feet long (the 'right-up-stick') which was braced upright by a line fastened to the junction of the snooks. The net itself was 20 feet long and fine meshed. Shrimping in the early days at Leigh was described by one of the old Leigh shrimpers as a matter of 'main strength and stupidness'. The trim-tram net was replaced in about 1830 by the Thames shrimp-net used by the Leigh men, and possibly developed by them from the trim-tram. Again the vital part was a beam of oak or elm between 8 and 12 feet long forming the bottom of the mouth of the net, a 'right-up-stick', placed vertically in a notch at the midpoint of the beam, which held a shorter 'thwart-stick' parallel with the beam and which formed the top of the net mouth. The shrimp-net was fastened to the towing warp by a pair of bridles. Later, the beam trawl was introduced, it is believed by the Gravesend and Barking fishermen, for deep-sea fishing and was adapted by the Leigh shrimpers about 1860. In effect, the beam was held clear of the sea bed by iron shoes at each end which ran along the bottom, while the lower mouth of the trawl ran loose protected by runners or chains. The beam was larger, 16 to 18 feet in length, and as the small 'pink-stern' boats were gradually replaced by the larger, decked bawleys, larger beam trawls – up to 30 feet – came into use.

Throughout the lower estuary shrimps were a major fishery. At Rochester in 1878 it was second only to the smelt fishery in importance. Twenty years later Murie (1903) reports average catches of 30 to 50 gallons, although some Leigh boats were fortunate enough to take 80 to 100 gallons a day. The record catch for a Gravesend bawley was about 120 gallons (Mansfield, 1922). A gallon contained some 3,000 shrimps. The annual catch must have been numbered in thousands of millions of shrimps.

Other fisheries mainly in the mouth of the river existed. Trawling for fish, mainly flatfish, although whiting and cod (in winter) would have been captured, was an important source of income, and the fishing fleet at Barking, no doubt, originally began by fishing the Thames mouth. In addition, oysters were cultivated at Leigh, and were occasionally dredged from natural beds, as off Garrison Point and along the Sheppey coast, clam digging and winkle collecting were locally practised, while the cockle beds at Leigh and below were heavily exploited. Some crabs and lobsters were caught, whelks and scallops were fished locally for a time and in the mouth of the river, whiteweed fishing (actually a plant-like colonial animal or hydroid) also provided a further source of income.

Throughout this chapter the constant theme for each of the major fisheries has been one of controlled exploitation providing employment for the fishing community (and the associated industries of boatbuilding, sail-making, rope- and net-making) and providing food for the population of the local communities and London. These ancillary industries were important; Venables (1874) records that a Fenchurch Street net-maker derived an annual income of £800 for salmon nets alone (this would have been in the beginning of the nineteenth century). Most of these fisheries were gravely diminished by pollution stemming from the development of London and interference with the river's flow. The up-stream fisheries for smelt, salmon, lampern, and shad were ruined, the middle-river fisheries for flounder, eel, and whitebait were seriously affected and those which continued had finally to move downstream to the less polluted waters of the mouth of the Thames. Even here, however, the fisheries were diminished and public health interests had made inadvisable the culture of filter feeding molluscs such as mussels and oysters. In short, pollution and other developments on the river had practically destroyed the most important fisheries in south-eastern England, fisheries which had supplied the London market since the first building of the city, and which had provided employment all along the river.

FOUR

THE RETURN OF FISH TO THE TIDAL THAMES

1900–50

The improvements to the river following the earlier attempts to treat London's sewage resulted in the return of some species of fish to the London Thames. As we have already seen (p. 50), smelt were captured at Blackwall, Putney, Isleworth, Kew, and Teddington in 1900, and for several years preceding, flounders also returned to these regions and whitebait were caught in the river at Greenwich. C. J. Cornish in his *Naturalist on the Thames* (1902) records the brief resurgence in fish life in the tidal Thames around the beginning of the twentieth century:

> Besides the estuary fish which naturally come *up* river, dace and roach began to come *down* into the tideway, and during the whole summer the lively little bleak swarmed around Chiswick Eyot. Later in the year the roach and dace were seen off Westminster, and several were caught below London Bridge, and in 1900 roach were seen and caught at Woolwich, but were soon poisoned and died. In August the delicate smelts suddenly reappeared at Putney, where they had not been seen in any number for many years. Later, in September, another migration of smelts passed right up the river. Many were caught at Isleworth and Kew, and finally they penetrated to the limit of the tideway at Teddington, and good baskets were made at Teddington Lock. . . .
>
> A few years ago hardly any fish were to be seen below Kew

> during the summer, and these were sickly and diseased. Last year they were in fine condition, and dace eagerly took the fly even on the lower reaches. Every flood-tide hundreds of 'rises' of dace, bleak, and roach were seen as the tide began to flow, or rather as the sea-water below pushed the land-water before it up the river. At high water little creeks, draw-docks, and boat-landings were crowded with healthy, hungry fish, and old riverside anglers, whose rods had been put away for years, caught them by dozens with the fly. Sixty dozen dace were taken, mainly with the fly, in a single creek, which for some years has produced little in the way of living creatures but waterside rats.

Unfortunately, Cornish's notes, which are often undocumented with dates and exact localities and are thus rather unsatisfactory, seem to be the last observations made by a naturalist on fishes in the tidal Thames. Occasional records exist in angling literature, but these were mostly written long after the years to which they refer. Thus, a letter published in *Angler's Mail* of 8 May 1974 from a Mr A. H. Comber of Canning Town recalled the period before the First World War, when there were eels and 'good fishing all the way along the River Lea to where it enters the Thames'. The same correspondent recalled collecting roach, tench, and eels from the dry dock of the Thames Iron Works Shipbuilding Company, and in 1903 when roach, perch, dace, and pike could be caught in the Victoria and Albert Docks.

The reference to fishes in the docks is, however, possibly slightly misleading if it is taken as an indication of the presence of fishes in the river adjacent to the docks. It seems certain that within the present century at least the docks have always held good stocks of fishes irrespective of the state of the river at their entrances. This is because they are operated to permit shipping to enter or leave through a series of locks, so that the ingress of polluted river water is always minimal and confined to ships entering the docks. Within the dock system ships do not discharge polluting substances to the water and mains services are available at the wharf. Apart from accidents and war damage therefore the docks have never been seriously affected by pollution. Even when the river was in exceedingly bad condition in the 1920s, the London Docks, which communicated with the river at the Pool of London, held abundant freshwater fishes. In 1921 Mr P. W. Horn, then Curator of the Whitechapel Museum, obtained roach, dace, bleak, gudgeon, perch, three-spined stickleback in numbers, with fewer bream, minnow, goldfish and pike being found. These fishes had succumbed to the temporarily unfavourable conditions in the dock during the exceptionally hot summer of that year (Horn, 1923). A very large number of fish were killed in the Surrey Commercial Docks that summer, an estimate of the total being in tons weight. The cause of these fish kills was attributed by Horn to deoxygenation of the water, the generation of foul gases from the bottom of the docks and a rise in salinity due to the river water penetrating the docks during the period of low freshwater flow which caused the proportion of salt in the river to rise above normal. All these causes might have contributed and at this distance in time it is not possible to confirm Horn's diagnosis, nor is it important to do so, for the example is quoted mainly to show how abundant fish life was in the docks.

Reports of fishes in the tidal Thames are unfortunately sparse for the period between the two world wars. There is no reason to doubt that freshwater fishes existed in abundance in the upper reaches of the tidal river below Richmond Lock, and down through Isleworth, Kew, and to Chiswick, for anglers fished these waters during this period. It has, however, proved impossible to establish where on the lower tideway fishes were still caught. The most important fishing communities on the lower Thames, at Gravesend and Leigh-on-Sea, were placed where they could turn to fish in the Thames mouth when pollution drove fish and shrimps from the river itself. The Gravesend fishery appears to have dwindled in the 1920s, while the Leigh men continued to earn a living catching shrimps and whitebait throughout that period, which suggests that there was a scarcity of animal life downstream to at least Gravesend even at that period.

Some observations on fishes in the upper tidal Thames in the 1920s have been received as a result of an appeal for information in the *Angler's Mail*. Major L. H. Osborne, RAMC (retired) in correspondence in March 1972, told me of fishing at Richmond in 1921 when roach and dace were in the river in thousands, with catches of fifty or more fish in an evening's fishing. Barbel were also plentiful there at this time. Between 1923 and 1948 Major Osborne fished from a punt below Teddington weir, and again found roach and dace in abundance, but after the severe winter of 1946–7 fish became scarce, the roach suffered from black spot (encysted flukes called *Cryptocotyle* the final host of which are gulls) and only gudgeon and a few perch were caught. The barbel had become less plentiful from about 1930. Major Osborne saw the decomposed body of a large fish at Petersham in 1921, which he firmly believed was that of a huchen or Danubian salmon, the introduction of which was discussed earlier (p. 69).

In the period 1929 to 1935 the late Mr J. E. Dandy, a noted angler as well as distinguished botanist and one-time Keeper of the Department of Botany at the British Museum (Natural History), fished the river near Kew, opposite Syon House. His observation that dace were especially abundant in that area and usually dominated his catches which were otherwise mainly roach is confirmed by other reports from that period. Mr Dandy's comments that the fish fauna between Teddington Lock and Richmond was undiminished then and had not been seriously affected by pollution is of interest, as it refers to the period before the war when there were the disastrous, and unauthorised sewage sludge discharges from Mogden Sewage Treatment Works.

In 1971 Mr John Burrett, a well-known angler and author of *Fishing the Lower Thames* (1960), gave me a summary of his fifty years of experience of angling in the tideway. He recalled fishing in the mouth of the River Wandle at Wandsworth around 1928 to catch small perch, while dace and bleak could be caught near Thorneycroft's boatyard at Chiswick. In 1930 he caught many roach near Kew Bridge, but between 1933 and 1955 he noted a marked reduction in the weed, molluscs, and leeches as well as fishes in this area. His experiences in the period 1929–34 showed that between Hammersmith and Putney the catches were mainly of bleak and dace, while roach occurred occasionally; bleak abounded and as many as thirty to fifty would be caught in a day. Near Kew Pier roach were more abundant and both they and the dace were larger than downstream. Nearby was a favoured site for dace, Mr Burrett remembers catching upwards of three score in a few hours, fishing in what was known locally as the 'Hot-Waters' (where a local laundry discharged hot soapy water to the river). This local warming apparently also encouraged perch and roach to accumulate in the area. Above this point Mr Burrett and his fellow anglers recognised several places where large fish congregated, and the general impression gained from his notes is that at this period the upper reaches of the tidal river contained large stocks of freshwater fishes.

In the 1920s fishes were evidently common in the river at Chiswick, as was borne out by the delightful note Professor Nicholas Polunin appended to a paper I wrote on the cleaner Thames and its fishes. He wrote,

> my brothers and I used on occasion to catch, with our hands or home-made devices, quite a number of freshwater fish – particularly Roach and Dace, but including occasional Trout and I think other species – in the pools left at low tide in the backwater of the

> Thames opposite our house on Chiswick Mall, in the shelter of Chiswick Eyot, the lowest island above London Bridge. From our front, riverside garden or upstairs bedroom windows we often watched Herons fishing successfully in those pools at low tide, while at high tide lunch-time anglers would catch an occasional Roach or Dace from the nearby banks or wharfs. . . . Sticklebacks abounded on the surface of the brimming marginal waters at high tide in summer, being frequently caught in dip-nets and confined to jam-jars by small boys who rudely called them 'pop-bellies'. (Polunin, N. in Wheeler, 1969).

An interesting incident during the Second World War has led to the suggestion that sticklebacks at least inhabited the tidal river in the vicinity of Chelsea. I first heard of this when Dr C. C. Hentschel, then retired from his former post of Head of the Botany and Zoology Department of Chelsea College, informed me that about 1944 he discovered that the static water tanks on the river embankment contained sticklebacks, as well as *Hydra* and aquatic insects. The local Fire Brigade confirmed that these tanks (which were a water reserve for wartime fire fighting) had been filled from the river. Therefore the sticklebacks must have been living in the river. This information was later confirmed by Dr Muriel Sutton of Chelsea College in 1969.

On the other hand I have been told by that well-known London angler and teacher, Mr Lewis Harris, that he had been instrumental in putting sticklebacks into static water tanks soon after they were filled as a means of controlling the numbers of mosquito larvae which found them an acceptable living space. So possibly there was another means of origin for the sticklebacks in London's static water tanks. The sudden and mysterious appearance of this species in a temporary pool of water was later observed in the outlet of a storm water discharge channel in the 1970s (see p. 158).

Our knowledge of the distribution and abundance of fishes in the tidal Thames is thus very limited in the critical period from 1900 to 1950. Such information as is available comes from isolated notes of this kind, and it can now be seen as regrettable that no naturalist made and published any observations on fish in the river. On the other hand it is entirely understandable that with the river in the condition it then was no one was inspired to investigate what would have been a largely absent fauna. The chemical condition of the river, with high levels of noxious effluents and little or or no dissolved oxygen, leaves little doubt that from

1914 there would have been very few fishes between Gravesend and Chelsea. It is not possible to verify this, and no doubt the situation changed seasonally, when high upland flows brought sweet water closer to London or when winter brought cooler weather and thus higher levels of dissolved oxygen, and also with the tides, as on the equinoctial spring tides which would have penetrated further upstream than the average tidal excursion. The situation could never have been static (which is one of the attractions of an estuary to the naturalist) but there is little doubt that the first half of the twentieth century saw a steady decline in the fish and the remainder of aquatic life, until a whole section of the Thames was virtually lifeless.

THE 1957 SURVEY

In 1957 I was invited by the officers of the London Natural History Society to compile an account of the fishes in the London area (which for the Society comprised a circle of twenty miles radius of St Paul's Cathedral, and thus encompassed a large part of the tidal Thames). The resulting paper was published in the centenary number of the *London Naturalist* (Wheeler, 1958). In compiling the information about the fishes of the tidal Thames, I made use of numerous sources; anglers who fished the river, electricity power station staff, river police, Port of London Authority employees, and dockers were invariably helpful with the information about fishes at their disposal. In most cases it was a depressingly uniform nil report, except for anglers. Anglers still fished the tideway in its upper reaches and reported that a moderately rich fish fauna still existed. From Teddington downstream to Kew fish, chiefly dace and roach, were caught in numbers, and it was not unknown for both species to be caught at Chiswick and Mortlake (Burrett, 1968). Mr Burrett amplified his published remarks later in correspondence and in the period 1953–7 he tells me that he caught roach, dace, bream, barbel, and a few chub at Brentford Dock. By about 1956 he had noticed that aquatic invertebrates had become more numerous than they had been formerly, and in a letter to *Angling Times* (7 June 1957) he told of fish caught at Brentford in September 1956 which included barbel, carp, 43 small roach, 7 dace, perch, and 2 eels. In May 1957 he saw numerous perch ranging from 8 to 15 inches in length and 2 eels. At this time he witnessed schools of dace from 4 to 9 inches in length trapped in a lagoon as the tide retreated.

These reports have supplemented my paper of 1958 but have not contradicted the conclusion that there was no evidence of fish life, with the exception of eels, in the tidal Thames within the metropolitan area. In the late 1950s the freshwater fish fauna was relatively rich downstream to Kew, and certainly thrived above Richmond Lock in the freshwater pounded from the extremes of tidal influence. Below Kew Bridge the only species to occur in numbers were roach and dace, which could be captured in the vicinity of both Chiswick and Hammersmith Bridges.

The presence of eels in the Thames, which was noted as the exception, proved to be a riddle to which the answer was only later revealed. Eels were known to occur in the tidal river at Richmond and at Chelsea, they were also found in the Lee at this period but their distribution was 'patchy' and they were not common. As the eel breeds in the sea and the young fish have to migrate upriver for larger, subadult eels to be found in freshwater, it seemed that this one species might have swum up the river, being able to tolerate the levels of pollution in the lower tidal river. An alternative means of entry into the Thames could have been through navigation canals connecting other English rivers to the Thames, but this could not be proved. The presence of eels in the tideway therefore led to the suggestion that 'it was probably the only species to pass regularly through the polluted lower Thames'. The evidence for this was circumstantial. How it was achieved was inexplicable in the absence of dissolved oxygen in the water. It is timely here to point out that eels do not 'take air at the surface' as so naively claimed by Harrison and Grant (1976). The partial explanation of the presence of eels has now been revealed. In 1946 the Thames was stocked with elvers apparently at the instigation of commercial fisheries, and there seems little doubt that these elvers spread downstream to give rise to the occasional eels captured in the tideway. As eels may stay in freshwater for up to twenty years before returning to the sea, this stocking could have resulted in eels occurring in the river until the mid-1960s. No doubt other stockings were attempted elsewhere and possibly there was some migration through canals.

Towards the sea the abundance of fishes in 1957 was less clear. No anglers fished the lower reaches of the river between London and Gravesend, nor did commercial fishing boats operate above Leigh-on-Sea, and so far as I could establish there were no casual observations from riverside workers on which one could draw. The then head of the Freshwater Fisheries Laboratory at the Ministry of Agriculture and Fisheries made enquiries on my behalf of his inspectorate concerning the presence of marine fishes in the mouth of the Thames. His reply (November 1957) that 'above Gravesend no sea fish are reported for the past twenty-five years. Before that there were a few flounders but they have disappeared owing to the growing pollution below London', seemed ample confirmation of the deductions made from the chemical condition of the river that at least as far downstream as Gravesend there were no fish in the late 1950s.

The summary of the investigations for the survey of London's fishes in 1957 still stands as valid. There was no established fish population between Kew Bridge to at least as far downstream as Gravesend. In the light of later experience it might be added that the fishless zone probably extended to the vicinity of Lower Hope Reach. There were thus some 69 km (43 miles) of river from which fish were excluded for the greater part of the year. It is, of course, not possible to be certain that no fish existed (absence is the most difficult feature to prove in ecology) in that region. Indeed, it is virtually certain that seasonal variation would have brought sea fishes upriver towards Gravesend in winter, while freshwater fishes would have been temporarily able to colonise further downstream than Kew Bridge with high upland river flows and lower temperatures. The often repeated statement that in 1957 there were no fishes (except for eels) between Richmond and Tilbury (cf. Harrison and Grant, 1976) thus requires some qualification, but is substantially true and stems from this survey which was conducted at the instigation of the London Natural History Society.

THE SURVEY OF 1967–73

The means by which it was possible to monitor the return of fishes to the tidal Thames was suggested by the happy chance of an engineer with an enquiring mind catching a fish which he did not recognise. In March 1964 a specimen of the tadpole-fish or lesser fork-beard, was caught on the cooling water screens of West Thurrock electricity generating station which was then being constructed. The specimen, which was 116 mm (4·7 in.) in body length, was brought by a Mr Colman, one of the engineers involved in building the new power station, to me for identification. The capture of this fish was at the time quite surprising, for West Thurrock, near Grays in Essex, lay well upstream of Gravesend, which only seven years previously had seemed to represent the lowest part of the fishless and anaerobic Thames. The tadpole-fish was also not well known; a small member of the cod family, it frequents rocky bottoms between Norway and the English Channel but is relatively rarely caught. Its capture in the

Thames seemed unusual in that it is rare in the southern North Sea where its normal rocky habitat is singularly sparse.

Mr Colman generously agreed to save any further fishes which were caught on the cooling water screens when they were being test run, and he was lent preservative, labels, and a receptacle in which to put the specimens. On 19 November 1964 he found a lampern, on 31 December another tadpole-fish, and on 15 January 1964 a sand goby and a stickleback were caught. The catches continued with smelt and greater pipefish (April 1966), John dory (November 1966), and a greater sandeel at some unknown date. Perhaps the most remarkable feature of Mr Colman's collection was the number of rather rare fishes he had found, for the lampern and John dory are, like the tadpole-fish, uncommon today in the southern North Sea. It was also interesting that as early as 1966 smelt and lampern had been found, for these, as we have seen, were fishes of great economic importance in the tidal Thames before it had become polluted.

Also in 1966, a newspaper reported that large numbers of roach had been captured on the cooling water screens at Fulham power station and that 20 to 30 lb of fish were being caught each tide (Marlborough, 1969). The value of this report was marred somewhat by the claim that a salmon had been caught there as well, which later enquiries led me to believe was either a large carp or a barbel! This report of roach being caught in numbers, which was subsequently confirmed by a visit to the power station, was of the greatest interest at that time for it showed that fishes were now being caught in an area which in 1957 was certainly fishless (Fulham power station is just downstream of Wandsworth Bridge and 11 km (7 miles) below Kew Bridge which had been the limit for substantial numbers of fish).

The coincidence that electricity generating stations at both Fulham and West Thurrock had been capturing fishes in the previously fishless zone of the Thames in the period 1965–6 suggested that such stations might prove to be successful fishing machines elsewhere. With the enthusiastic support of Mr John Wilson and Mr K. M. Gammon, of the Central Electricity Generating Board's Station Planning Branch, arrangements were made to set up a series of collecting points at Thames-side power stations where containers with preservative could be sited into which any fishes caught could be placed. The original list of generating stations was compiled with two factors in mind; first, they needed to have the right kind of circulating water intake screens (technically, moving band screens); second, a wide coverage of the conditions in the river was

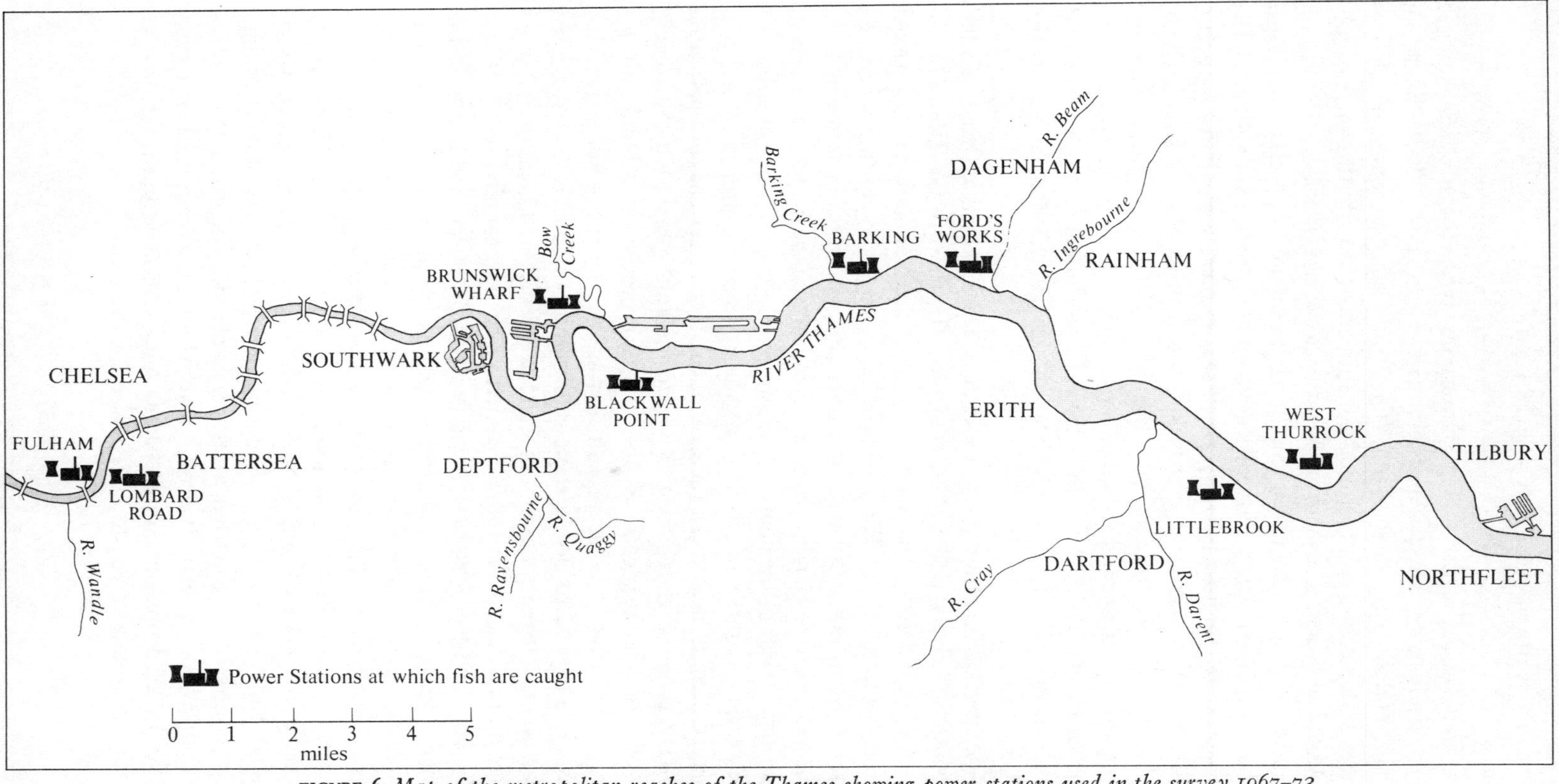

FIGURE 6 *Map of the metropolitan reaches of the Thames showing power stations used in the survey 1967–73*

required. Initially, the power stations involved in the scheme ranged from Fulham to West Thurrock, with Brunswick Wharf, Blackwall Point, and Barking in between. They thus sampled water which was fresh and relatively unpolluted at Fulham, through the increasingly polluted and saline reaches at Brunswick Wharf, Blackwall Point, and Barking, to the nearly marine conditions but moderate pollution at West Thurrock. Later, other electricity generating stations which joined in the survey were Lombard Road, Battersea, (which was almost opposite Fulham power station) in February 1969, Littlebrook, Dartford, (lying a little upstream from West Thurrock) in February 1970, and the company station at Ford's Works, Dagenham, in May 1972. As it evolved, the series of collecting stations gave a broad picture of the return of fishes to the tidal Thames which would have been unobtainable by any other means.

The mechanics of using electricity generating stations as 'fixed engines' for fishing should perhaps be explained. In the process of producing electricity power stations use water to cool the steam after it has passed through the turbines. The surplus heat is taken up by this water and discharged in one of two ways. Either the station has a series of huge milk-churn-shaped cooling towers through which the heated water is run, the surplus heat in the shape of steam being lost to the atmosphere, or it uses large quantities of river or sea water which is pumped in and discharged slightly heated. In the case of the former system relatively little water is required to replace that lost to the atmosphere, but the cooling towers are often regarded as unsightly, are expensive to build and use large areas of land, and as a result they are rarely used in new generating stations today. Most power stations are built beside rivers or on the sea coast where adequate supplies of water are available. This is drawn in by huge pumps, usually through channels or tunnels running out well into the river. In the process of drawing in water any suspended matter in the water is sucked in also, and arrangements must be made to remove this before the water is circulated into the power station. In many stations this is done quite simply by having a wide-meshed grid on the intake entrance (which will stop objects such as timber or oil drums from entering the system) and a finer-meshed moving band screen which catches the smaller material. These fine-meshed screens, when working, move continually, rather like a vertical escalator, and have narrow ledges on each section to retain the solid material which, once it reaches the top of the band, drops off into a channel and is washed into a pit or over another retaining screen to be removed for disposal.

Any fishes which are swept into the intake channel with the cooling water will eventually be carried onto these screens along with the rubbish of the river, and it is on the screens, in the channels from them, or in the rubbish pit into which these discharge that they can be picked out.

Because it was clearly impossible to sit over the screens at all the eight stations, or even the original five, so that all fishes caught were retained for examination, arrangements were made with the superintendent of each station for the staff who were responsible for cleaning the screens to save any fish they found, put them in preservative, with a label to indicate date of capture. This led to variation between power stations, because the preserving of fishes depended entirely on the good will of the staff who undertook this additional work in addition to their official duties. It is appropriate here to pay tribute to the care that they took to preserve fishes for study, and the enthusiasm which they and other power-station staff showed was a constant inspiration.

Unfortunately, the numbers of fishes caught at both Fulham and West Thurrock in the later years of the survey caused problems and it was possible only to save a sample of the fish recovered from the trash pits. Additionally, the gulls proved serious competitors for small fishes at West Thurrock, and specimens were often snatched from the upper part of the screens by them. Upstream, at Fulham, the station cats sometimes seized choice specimens before they could be preserved for scientific examination!

These vagaries of collection probably had little effect on the total picture which emerged as the fish returned to the river. Other factors influenced the results more seriously. Thus some of the generating stations, chiefly the smaller and older ones such as Lombard Road, did not run continuously, as did the modern stations such as West Thurrock which supplied 'base loads' to the national grid. Pumps and screens periodically required to be serviced or dismantled and were sometimes not used for long periods of time. Finally; the quantity of water required by the power station varied with the requirements of each even when they were all working supplying electricity at peak periods. It was therefore never possible to make direct comparisions between the catches of fish at different stations, nor was it possible to make predictions of the abundance of a species in the river other than in the most general manner.

The survey did, however, provide evidence of the gradual increase in the number of species in the river and the improvement in the fauna as a whole. It was essentially qualitative not quantitative.

It is here worth pointing out that some workers using power stations as a means of collecting fishes have attempted to make correlations between the number of fish caught and the known volume of water pumped in from the river. Such a relationship is less accurate than may be apparent, as the position of the intake affects the composition of the fish fauna drawn in by the pumps, and the velocity of the pumped water has a strong bearing on the species captured and the size range of the individuals of the same species. An example of the latter is the large number of young bass caught at West Thurrock, compared with adults, which is not in direct proportion with their numbers in the river but is really a reflection of swimming ability correlated with size of fish; that is, the large fish can escape the inflow current. Attempts at quantitative sampling using power station intakes need to be regarded with suspicion.

The power station survey was supplemented by other means of collecting fishes. Several experimental fishing competitions were organised by the Greater London Council in which teams of anglers fished in the reaches in central London. Their catches were examined and helped to fill in some gaps in the picture. With the assistance of the Port of London Authority's staff at Kew a series of nettings of the upper river were made, with remarkable results, and towards the mouth of the river beam trawling from the Ministry of Agriculture, Fisheries and Food's research vessel *Tellina* helped complete the picture below Gravesend and on the sludge dumping grounds in the estuary. The results of these supplementary surveys all combined to present an overall picture of the fishes when they and other aquatic life were returning to the tidal Thames.

Between August 1967 and December 1973 a total of 68 species were captured at the generating stations on the river (other species, to make a total of 72, were caught by other means within the tidal reaches). Of this total 18 were freshwater species, 43 were of marine origin, and 11 were euryhaline, capable of living in both saltwater and freshwater. In a sense this last group were the most interesting because no fewer than 6 were migratory species some of which were known to have swum up the river into freshwater from the sea and thus established that the whole length of the tidal river was in a condition suitable to support fish life.

Analysing the catches in detail is best attempted by taking the stations as pairs and examining their catch in terms of numbers of species and relating it to their position on the river. This is supplemented by Figure 7, which shows the proportions of marine or freshwater species which each pair of stations caught in relation to salinity of the water. To avoid ambiguity the pairs of stations are referred to by their approximate location on the river, not by their names, viz. Fulham and Lombard

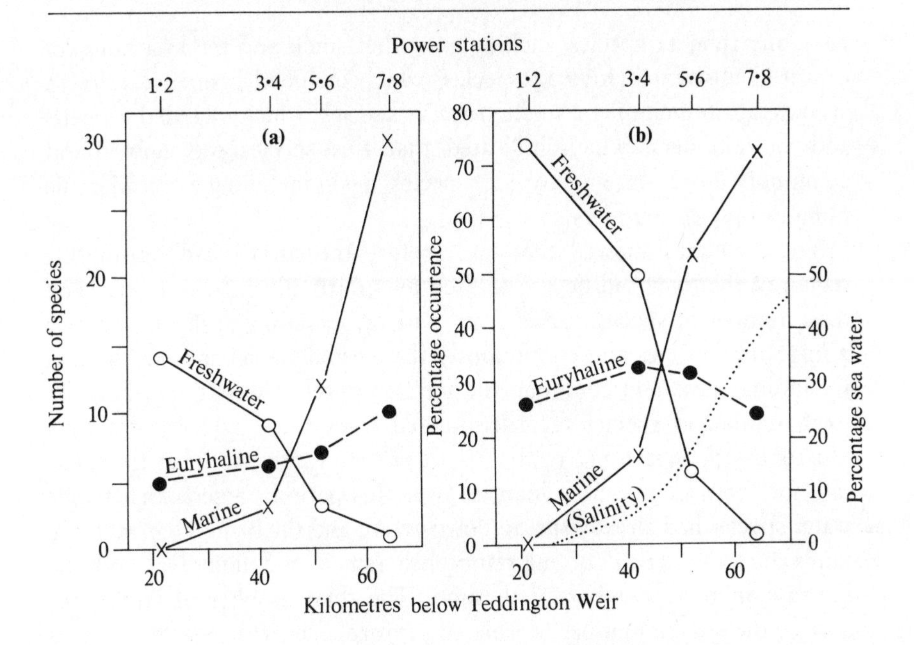

FIGURE 7 *Analysis of species in the tidal Thames at four pairs of power stations in 1967–73. This shows (a) numbers and (b) proportions of freshwater, euryhaline, and marine species; (b) also shows the approximate half-tide salinity (as % sea-water) for average flow conditions 1967–73. Pair 1 and 2 Fulham and Lombard Road; pair 3 and 4 Brunswick Wharf and Blackwall Point; pair 5 and 6 Barking and Ford's Works; pair 7 and 8 Littlebrook and West Thurrock*

Road are referred to as Wandsworth Bridge, Brunswick Wharf and Blackwall Point are referred to as Blackwall Tunnel, Barking and Ford's Works, Dagenham, are referred to as the Outfalls, and the generating stations at West Thurrock and Littlebrook as the Dartford Tunnel.

WANDSWORTH BRIDGE AREA The two generating stations in this vicinity (about 9·7 km above London Bridge) caught a very large number of freshwater fishes and an increasing number of migratory fishes which had penetrated upstream. Chronologically their joint catches were as follows: 1967 9 species all freshwater or euryhaline; 1968 13 species, including a smelt, the first migratory fish to be captured upstream of

London; 1969 11 species, including another smelt and the first flounder captured upstream; 1970 14 species, which included 2 smelts; 1971 13 species, again including a smelt; 1972 11 species, which included 2 smelts and 74 flounders, which indicated that this species was now found commonly upstream; and 1973 11 species, again including a considerable number (45) of flounders.

These catches showed that even before the survey had begun this region of the river had been colonised by a fairly diverse fish fauna. The total number of species varied from 9 to 14, including at different times 2 migratory species and continuously the euryhaline eel and stickleback. Excluding these and confining the total to purely freshwater fishes, the total number of species recorded ranged from 7 to 11 (1967 – 7; 1968 10; 1969 – 7; 1970 – 11; 1971 – 10; 1972 – 7; 1973 – 8). That the totals did not show a continuing increase over this period suggests that freshwater species had already spread downstream and the major change in the fauna was the arrival of migratory fish (smelt and flounder) and the increase in their numbers with time. The total number of freshwater species included a number of isolated captures each year, viz. barbel and tench in 1967 and 1968, rudd in 1971, chub in 1968 and 1970 (2), but the catches throughout this period were dominated by the roach, bream, bleak, and perch, while pike occurred regularly but in small numbers after 1969 and dace only from 1971. Curiously, the ruffe, which occurred in the samples in 1967–71, was not caught in later years.

In this region the most significant catches were undoubtedly those of the migratory smelt and flounder. The smelts first caught were large, mature fishes between 150 and 175 mm (6 to 6¾ in.) in body length, but in early 1972 a small specimen, only 67 mm (2·6 in.) long was caught. This was the first suggestion that the species might again now breed in the Thames itself, as opposed to the tributaries in the outer estuary where the species was known to breed, and presumably always had done so. The flounder first caught here in 1969 was small, only 87 mm (3½ in.) long, but in 1972, when many were taken, it was obvious that young fish had migrated up river in large numbers during the preceding year.

BLACKWALL TUNNEL AREA The pair of generating stations in this region, at either end of Bugsby's Reach and just downstream of the tunnel (around 11·3 km below London Bridge), produced some of the most interesting results, for they are sited well within the industrial reaches of the river and closer to the sewage treatment works outfalls at Beckton and Crossness. The River Lee discharges into the Thames close to Brunswick

Wharf and at the outset of the survey was seriously polluted in its lower reaches. In this region the salinity is low (around 5 per cent sea water at half-tide conditions and for average freshwater flow), but it is marked enough to have an affect on the fauna. Here again, freshwater fishes were dominant, the euryhaline species were almost the same as at the upstream stations, but a few wholly marine fishes were also caught.

Chronologically, the catches by power stations in this area were as follows: 1967 7 species were caught, all freshwater or euryhaline in nature; 1968 15 species were taken, including the migratory smelt and the first bass to be reported so far upstream, as well as 2 marine species, a sandeel and a sand goby; 1969 yielded 13 species, which included 3 smelts, and the following marine species – 16 sprats, 1 tadpole-fish, 1 sandeel and 2 sand gobies; 1970 14 species were caught, including 9 smelts and 1 flounder, a sprat and a sandeel; in 1971 15 species were caught, including smelts (3), sand gobies (5), thin-lipped grey mullet (1), and sprats (12); in 1972 only 11 species were captured, including 2 smelts, 28 flounders, and 1 tub gurnard; and 1973 7 species were recorded amongst which the most notable were 20 flounders.

The survey here showed that the fish fauna was well on the way to recovery in this area even as early as 1967 but that the situation continued to improve through to 1972. The numbers of species caught rose from 7 in 1967 to 15, 13, 14 and 15 in successive years, and although the total number of fish available is not statistically significant, the annual catch did rise from 52 in 1967 to 128, 97, 69, and 142 between 1968 and 1971. The purely freshwater species were always abundant, most notably the roach and the bream, but from 1968 onwards perch became distinctly more common, as did bleak from 1969, and dace from 1971. The increase in numbers of these 3 species, which all seem to have somewhat more rigorous demands for quality of water than roach or bream, suggests that there was a positive immigration of these species as the river improved. The return of these freshwater species was apparently reflected in the increase in numbers of smelt commencing in 1968, and flounders from 1972, while truly marine species were first recorded in numbers around 1969 (sprat) and 1968 (sand goby).

In addition to the abundant freshwater species caught a number of interesting less-common species were recorded. It was always a matter of surprise that the majority of the brown trout caught by power-station sampling were taken in this area – 1968 (2), 1971 (1), 1972 (1) – and indeed the first specimens were examined very carefully to ensure that they had been fresh when first caught and were not frozen specimens jettisoned from a docking ship's freezer. Eventually it was concluded that all were native Thames trout. Almost equally curious, if for different reasons, were the number of goldfish captured here; 1967 (2), 1968 (2), 1969 (1), 1970 (2), 1971 (3). This species was also captured in the Wandsworth Bridge and the Outfalls areas and these captures suggested that the goldfish was well distributed in the river. Amongst the other rather unusual fishes found in this region were pike 1967 (1), 1968 (2), and 1970 (1), tench 1970 (3), chub 1971 (1), and ruffe 1967 (1) and 1969 (1).

Other euryhaline fishes, notably the eel and the stickleback, remained relatively consistently common during the five-year period. Thus, the eel was recorded in 1967 (2), 1968 (5), 1969 (1), 1970 (3), 1971 (19), 1972 (3), and 1973 (2), although these represent minimal figures as eels are edible and highly favoured by some and, no doubt, not all that were caught were preserved for science. These factors do not, however, apply to the stickleback, which was recorded as follows; 1967 (7), 1968 (30), 1969 (42), 1970 (13), 1971 (18), 1972 (15), and 1973 (2). Such consistent captures suggest that there were resident populations of both species in this area and that the stickleback, at least, was present in some numbers.

THE OUTFALLS AREA The two power stations in this area are Barking, a CEGB station, and Ford's Works at Dagenham. Both are sited on the north bank in Essex, Barking just downstream of the mouth of the Roding at Barking Creek and close to the Northern Outfall at Beckton, and Ford's power plant just downstream of the Southern Outfall at Crossness and by the confluence of the River Beam and the Thames. Because of the proximity of the outfalls, this region of the river had been seriously polluted for many years and indeed was still at the bottom of the oxygen sag curve in the early 1970s. In addition, there is a considerable amount of industry along the river here and this, and the upstream generating stations, as well as the treated sewage effluent, combine to elevate the river water temperature. It is clear, therefore, that the analysis of the results of the survey here were of major interest. Unfortunately, the Ford Motor Company power station was not involved in the survey until May 1972, so the earlier and critical years were not covered as fully as the later years.

The catches in this region were until 1971 always much poorer both in number of species and in number of individuals than either upstream or downstream. The totals were, 1967 (5), 1968 (9), 1969 (6), 1970 (1), 1971 (3), 1972 and 1973 (17); the increase in the last two years was due to the contribution made by the Ford's Works plant. As might have been predicted, the fauna showed a reduction in the number of purely freshwater species and an increase in the number of marine species, while the non-migratory euryhaline species remained about the same. Freshwater species comprised the roach, captured in 1967 (5), 1968 (6), 1969 (1), and 1970 (2), the bream, captured in 1968, 1969, and 1971 (1 each year), pike 1969 (1), carp 1968 (1), gudgeon 1968 (1), goldfish 1967 (1), crucian carp, 1973 (1), dace 1972 (1), and perch 1967 and 1972 (1 each year). Comparison of the catches of freshwater fishes in the Blackwall Tunnel and Wandsworth Bridge areas shows how attenuated the fauna was near the sewage outfalls.

Marine fishes were not abundant either, but their occurrence was of great significance here in that they demonstrated that fish could survive the passage from the sea through the gradually worsening conditions to the worst part of the river. The first capture was of one sprat in 1967, a species which occurred again in 1968, 1971 (both single fish), 1972 (5), and 1973 (.166). Bib or pouting were caught in 1968 (2), 1969, 1972 and 1973 (1 each), but the most surprising capture was of two haddock in 1969, a year which saw a large number of this species in the southern North Sea and Thames estuary. Sandeels were caught in 1968 (1) and 1972 (2), and Raitt's sandeel in 1973; tub gurnards were caught in 1972 and 1973 (2 in each year); scad or horse mackerel in 1972 (3) and 1973 (1); in 1973 two species of pipefish (5 fish), transparent goby (1), sole (1), dab (2) were all caught, as was a single anchovy in 1971. The most striking result in terms of numbers was the capture at Ford's Works of the estuarine sand goby in 1972 (33) and 1973 (863). Although this species was also captured in the Blackwall Tunnel area and one was picked up at Fulham in January 1969, it was nowhere common. That it appeared in such numbers once the Ford's Works plant was saving fishes may have been due to colonisation in large numbers in the years 1972 and 1973, but is more probably due to the siting of the power plant's intake in the river, and its sudden occurrence can be attributed to this circumstance.

Euryhaline fishes in the Outfalls area were, as might be expected, frequent in their occurrence there. One such was the common goby, which was caught, again at Ford's in 1972 (70) and 1973 (77). Possibly this local abundance was again due to the position of the station intake close to the confluence of the River Beam with the Thames. This goby was known to be common in the upstream Beam, as it was, and is, in other Essex tributaries such as the Mardyke and the Ingrebourne. The related sand goby has already been mentioned. Sticklebacks were relatively common at all times in this area, the total catch being 1967 (3), 1968 (3), 1969 (1), 1972 (60), and 1973 (6). Some migratory species also occurred during the whole period, the eel notably, 1968 (12), 1972 (11), and 1973 (4), while the flounder was caught there only in 1972 (12) and 1973 (1), and single smelts in the same years.

Overall, in this region the fish fauna was clearly impoverished, although by 1972 it showed signs of becoming richer both in species and numbers of fish. Clearly, the relatively low levels of dissolved oxygen in the water was delaying recolonisation, but the water was nevertheless suitable at times at least for marine fishes to penetrate and for migratory species to pass through. This most sensitive region began to recover from the long years of pollution during the survey and the return of fish here marked a milestone in the restoration of the river.

THE DARTFORD TUNNEL AREA Two power stations contributed to the study of fishes in this region, West Thurrock near Grays, Essex, and Littlebrook, Dartford, Kent. They lie respectively downstream and upstream of the Dartford Tunnel, and some 54 and 52 km (21 and 20 miles) below London Bridge. The contribution of West Thurrock generating station in catching some of the first fishes caught in the cleaner Thames in 1964–5 has already been mentioned. In this region the water is saline, at around half-tide and in average flow conditions being about 37 per cent seawater. Despite this, from all accounts, during the period of severe pollution conditions in this region were bad, dissolved oxygen levels in the summer being below 10 per cent of saturation and in the 1957–8 survey there was no evidence of fish life, or significant aquatic life at all.

By the time regular sampling was started (November 1967) at West Thurrock the fish fauna was already in the process of becoming well established. For example, 126 sprats were collected in a short period on 10 January 1968, and some 26 species were recorded in the first two months of collections. The number of species recorded in this region was high (it totalled 51 by the end of 1973) and sometimes the number of individual fish was so large that it is difficult to give a brief survey here, although fuller details are listed under the species accounts. Here the results are given in a general form with emphasis on the most abundant species or those which are of especial interest in some way or another.

Throughout the period 1967–73 the species which dominated in the catches were young clupeids, particularly sprats and herring. The sand goby was common in most samples, increasing in number from the winter of 1969–70 and continuing at a moderate level during each winter. Young gadoid fishes were present throughout, but from 1967 to 1969 only a few whiting and occasional pouting were caught other than singly. From the winter of 1969–70 both species began to increase, and occurred in numbers between 10 and 50 in the wintertime samples each year. Flatfishes of various kinds had also been recorded from the earliest samples and several flounders were taken in the first two months. Strangely, they were exceeded in numbers by young soles, which might have been expected to have been the rarer species, and 24 small soles were preserved in the January–April 1968 samples. This species remained a relatively common fish in the region in the winter-spring samples through to 1973, and later. Young plaice were also present but did not occur in numbers until 1971 when they were caught at both stations. The dab likewise was first caught early on as single fish, but in the autumn of 1970 increased in numbers and was common in the winter of 1971. Although evidently present in the river thereafter, it was not exceptionally abundant. The interest of the increase in numbers of the flatfishes, and to a lesser extent the sand goby, is that these are bottom-living fishes which must be subjected to the quality of the water near the river bed, and they are not, like the sprat, herring, whiting, and bass mid-water or surface-living fish which can swim in the well-oxygenated surface layers of the water. They are therefore in a sense better indicators of the improved quality of the river than many other species. The relative abundance of young bass was a surprising feature of the collections in this area, from the 12 specimens that were caught in November 1967 to January 1968, to the 30 of December 1969. It might have been expected that this species would shun an area of relatively low dissolved oxygen level but it was proved to be relatively common throughout the period 1967 to 1973, especially in the winter months.

Whiting were rare in this region during the early part of the survey, although the species was first recorded in the winter of 1967–8, but the following winter their numbers increased, and by 1973 numerous specimens were caught at each collection. For example, 26 specimens were preserved at Littlebrook generating station during the month of December 1971. The bib or pouting, which like the whiting is a member of the cod family, was occasional in the catches in the Dartford Tunnel region, from its first occurrences in November 1967 and January 1968 throughout the survey, but numbers increased notably from the winter of 1970–1 and continued at a high level thereafter between the months of November and April. Other members of the cod family caught here included young cod, haddock, tadpole-fish, and five-bearded rockling.

As with the Outfalls region, the euryhaline and migratory species were much in evidence; flounders and bass have already been mentioned, smelt, eels, and sticklebacks all occurred there as well from the earliest days of the survey (1967–8). However, their numbers were relatively small at first, although both the former became increasingly common as the years progressed. The smelt was first caught here in the winter and spring of 1968 (4 specimens) and each year thereafter in small numbers. At this period all those caught between December and April were sexually mature and ripe and were evidently potential spawning fish. The eel, which occurred in numbers further upstream, was comparatively uncommon, and although it was first caught as early as November 1967 only single large specimens were captured each winter until the spring of 1972 when several (3–4 fish) were caught in each monthly sample. The most significant captures, however, were of elvers, the young eels which had just completed their transatlantic migration from the spawning grounds. The first captures were made in early April 1968 (Robin Huddart, personal communication) and continued in small numbers until 24 May 1968. The average length of these elvers was 66·5 mm (2½ in.) These small fish were the first elvers to be reported entering the river this century, the forerunners of the now abundant elver migration up the Thames each spring, which recalls the celebrated eel-fares of previous centuries. The three-spined stickleback was not abundant in the early part of the survey. It was first recorded in April 1968 and December 1969 as single fish, but several were caught in the winter of 1969–70, and it became increasingly regular in occurrence although never abundant in succeeding years.

Other less common migratory fishes also occurred. A lamprey was caught in November 1967 but the species was not recorded again until the summer of 1973. The lampern likewise was caught on two widely separated occasions in 1970 and 1973. Twaite shads were caught as single fish in the summers of 1968 and 1973 and the winter of 1969. These species presumably owed their occurrence to chance penetration of the estuary by small schools or individual fish entering from the North Sea. Although their occurrence was of interest, it was significant only in the context that they had survived in the river water long enough to enter the Long Reach area.

Equally of interest but of no greater significance was the occurrence of truly freshwater fishes apparently alive and well in this saline area. They were never numerous but over the years 1967–73 the following species were caught; brown trout (1971), pike (1971), carp (winter 1969–70), gudgeon (1970), bream (1970), roach (1967), and perch (1970). Each species was represented by a single fish, and three of them (gudgeon, bream, and perch) all occurred in a single sample, February to May 1970, which suggested that they had perhaps been carried by high freshwater flows downstream, or more probably out of a lower tributary such as the Darent or the Cray.

As one might have expected, the salinity in this region had a major effect on the fish fauna. Marine species far outnumbered all others as, of course, they do in the European fauna in general. Freshwater species were virtually non-existent and possibly owed their occasional presence to local cataclysms. The euryhaline and migratory species were present in some numbers (11 species), all but 3 of which were also encountered at sampling stations upstream. The 3 exceptions, lamprey, lampern, and twaite shad were only rarely encountered. If salinity had been the only limiting factor in the distribution of this group of fishes, one would have expected all the species to occur throughout the tidal river (as historical evidence suggests they had done before the river was polluted). The condition of the water in the period 1967–73, and the long history of gross pollution, were then clearly affecting the distribution and occurrence of migratory fishes within the river.

That the low levels of dissolved oxygen in the water between June and September at this period and in this region had an effect on the fish life was demonstrated by the very few specimens that were caught at the generating stations. In 1968 only 9 specimens (6 species) were present in the samples between April and October; no fish were found before November 1969, and in 1970 only 7 specimens (6 species) occurred. The pattern was similar in 1971, but in 1972 and 1973 rather more specimens and species were present (20 specimens, 15 species in the latter year). The data do not, however, support further detailed analysis due to the methods by which the fish were collected (for one reason demand for electricity is lower in Britain at this period and the plant may therefore not be used at full capacity). The conclusions from this work are therefore subjective but they do show that in the period 1967–73 an increasingly rich fauna became established in the area, that there was a distinct increase in the number of individual fish caught, and that in the second and third quarters of the year there were few fish present in the river, due presumably to the lower levels of dissolved oxygen, but that absence at these seasons in 1967–8 had given way to sparsity by 1972–3.

OTHER SOURCES OF INFORMATION

The survey of tidal Thames fishes using electricity generating stations had obvious limitations as to the amount of information one could derive from the fish obtained. Notably they produced detailed information from a series of fixed points along the river, but this led to an increased awareness of the sparsity of knowledge about the fish fauna above and below the extremes of the sampling points as well as in between them. For this reason every effort was made to obtain information about fish in the river by other means.

Below the Dartford Tunnel area it was possible to trawl using a beam trawl with shrimp netting from the research vessel *Tellina* of the Ministry of Agriculture, Fisheries and Food (by invitation of the Naturalists-in-Charge – Dr R. G. D. Shelton and Mr J. P. Riley) in August 1970 and May 1971. In August 1970 a series of stations were worked on the Yantlet and Blyth Sands and Mucking Flats as far up as Coalhouse Point, these areas lying between 48 and 54 km (30 and 40 miles) below London Bridge in Lower Hope and Sea Reaches. The cruise, selected deliberately as the least favourable period of the year for the dissolved oxygen levels in the water, was productive and filled a lacuna in our knowledge of the fauna between Tilbury and Southend-on-Sea where it was known some commercial fishing and much angling was successful.

In Lower Hope Reach along the edge of the Mucking Flats the catch in 5–10 m depth comprised two small soles, many sand gobies, a large number of brown shrimps, many shore crabs and swimming crabs, *Macropipus holsatus*, numerous amphipod crustaceans, *Gammarus locusta*. Eastwards, towards the sea, the fauna was richer. On the West Blyth Sands (4–6 m of water) young flounders and soles were caught, as were numerous sand gobies, and a young herring. In deeper water, thus towards the main shipping channel, larger flounders and large numbers of brown shrimps and the same swimming crab were captured. On the Yantlet Sands flatfishes were again numerous, and here flounders, soles, plaice, and dab were caught in some numbers. Sand gobies were very common, while young whiting, garfish, five-bearded rockling, and Nilsson's pipefish were also caught. Here too the crustaceans caught included brown shrimps, shore crabs, the amphipod *Gammarus locusta*, and the opossum shrimp, *Neomysis integer*.

In May 1971 trawling was conducted further out to sea in and along the Barrow and Black Deeps, as well as along the submerged edge of the West Barrow, an elongate sand bank exposed at low tide, which forms the north-west border of the Barrow Deep. The interest in establishing what fish lived in this region was that these two deeps had in turn served as the dumping grounds of London's treated sewage sludge. The results exceeded all expectations, for as far as the fish were concerned there was little sign of a diminution in the fauna, although the most productive stations worked were in the shallower water at the edges of the deeps where possibly the tidal currents had scoured the sludge away from the bottom. However, fish and invertebrates were found in the depths of the channels, amongst the debris of the sludge.

In these channels 16 species of fish were found, of which the sand goby was by far the most abundant. Less numerous, but nevertheless still abundant, were Nilsson's pipefish, transparent goby, weever, hooknose, dab, and plaice. Amongst the unusual species collected here were a scaldfish and a small spurdog, the latter the only record of this small shark in the present work on the Thames estuary.

The limitations of these trawling surveys were considerable. The beam trawl is of high efficiency when operated by skilled fishermen, but it only fishes a narrow path on the sea bed (as narrow as the width of the beam used), and as the net was of fine mesh it was towed slowly. This certainly meant that large fish were able to avoid capture by swimming away from the mouth of the net, and indeed most of the fish caught were small, bottom-living species or young specimens of larger species.

These surveys provided interesting data when compared with and supplemented by the information from the power station survey in the mouth of the river. They also supplemented the very considerable information available from my trawling surveys in the mouth of the River Crouch and to seaward of Foulness from 1959–66, and in the mouth of the River Blackwater in 1960–1, both integral parts of the Thames Estuary.

Upstream, above the Wandsworth Bridge region, rather little was known about the fish in the river until one reached Richmond where considerable angling took place. For this reason during the summers of 1971 and 1972, with the assistance of the Port of London Authority's launch and staff at Kew, a series of seine nettings were made between Isleworth Eyot and Barnes Bridge, approximately 24 and 17 km (15 and 11 miles) above London Bridge. Using a 100 m seine net with a fine mesh, which was efficient in capturing small species and small specimens of large species if not large fish, a considerable number of fish were caught. Between Isleworth Eyot and Mortlake the most abundant species was the bleak, with nearly 100 fish in each haul. The three-spined stickleback was next most abundant (17 fish per haul), and the roach and gudgeon both averaged 10 fish per haul. Dace were less common, and few specimens of bream, loach, and minnow were caught; perhaps the most interesting capture was of 5 flounders at Isleworth Eyot in July 1972, probably the first to have been netted upstream of London since the commercial netsmen had ceased working the river in the previous century.

Netting in the Petersham Meadows area, upstream of Richmond Lock in both 1972 and 1973 produced large numbers of bleak (the largest haul was 403 fish in a single sweep), dace, roach, and gudgeon. Barbel, loach, pike, minnow, and perch were less common. Here, in the summer of 1973 over 50 flounders were caught in one day, in a sweep which covered the muddy bank upstream of the Eyot. This area of soft bottom was where most of the abundant gudgeon were captured. Apart from the capture of the flounders, which was the earliest evidence of the species above Richmond Lock, all the other species were known from anglers' and other records to inhabit this part of the river.

Since the formation of the Thames Water Authority a small team of biologists, led by Mr M. Andrews, has continued biological monitoring of the tidal river. This has produced more information on the continued increase in the diversity of the fauna in the vicinity of the Dartford Tunnel (mostly by sampling at West Thurrock Generating Station), and in the, often dramatic, seasonal changes in abundance in the composition of the fauna. Mr Andrews has made some information on the catches at West Thurrock and other Thames-side power stations available to me. He has also published (Andrews, 1977) an account of the changes in the tidal Thames fauna during the drought of the summer of 1976, which showed that such marine species as the sprat and sand goby were found as far up as Chelsea Reach. Their presence there, and that of numerous marine invertebrates, such as the oppossum shrimp, *Neomysis integer*, the prawn, *Palaemon longirostris*, and the brown shrimp, *Crangon crangon*, was due to the increased penetration of salt water above London Bridge as the freshwater flow over Teddington Weir declined with the low rainfall and the continued abstraction of drinking water upstream.

In the following chapters information available on each species of fish recorded from the estuary is given. Because they will be more convenient to use these data are presented in chapters on freshwater fishes, estuarine and migratory fishes, and marine fishes, with cross-references where a species might be expected to appear under two or more headings. The systematic arrangement, nomenclature, and general notes on biology derive from the most recent authoritative source on European fishes (Wheeler, 1978a).

FIVE

FRESHWATER FISHES

The fishes treated in this chapter are those that always live in freshwater and which cannot tolerate any excessive degree of salinity. This physiological distinction sets them apart from the estuarine or salt-tolerant species such as the lamprey, lampern, eel, the twaite and allis shads, salmon, trout, and rainbow trout, and the three-spined sticklebacks, all of which are included in the next chapter.

PIKE FAMILY

Pike, *Esox lucius*, a fish which is widespread across the whole of northern Europe (north of the Pyrenees and Alps) and Asia, extending into North America where it is known as the northern pike. It is a solitary fish which is predatory from early life, and from a length of about 30 cm its diet is almost entirely fishes. Typically it stalks its prey from the concealment of weed beds or bankside vegetation, attacking with a sudden bold rush. It identifies its prey by sight, and possibly by vibrations reaching the sensory pores on the head and body. It is unusual for the pike to be present in great numbers in any natural water on account of its life style.

Its occurrence in the tidal Thames was interesting because of the lack there of suitable vegetation in which to hide, and because of the heavily silted water, which must reduce vision to a very low level. Nevertheless pike were caught regularly but in small numbers from the outset of the survey at the Blackwall Tunnel region (1 – 1967; 2 – 1968; 1 – 1970), a single specimen was taken at Barking in late 1969, and at the Wandsworth Bridge area in early 1969 – here the numbers of fish caught were higher (1 – 1970; 3 – 1971; 1 – 1972; 3 – 1973). These fish were young and relatively small, at the Blackwall Tunnel area they ranged from 270 to 570 mm (10·5–22·5 in.), at Fulham from 220 to 600 mm (8·7–23·6 in.), although all but one were greater than 470 mm (18·5 in.), the mean length being 500 mm (20 in.). The largest Fulham fish weighed 3·03 kg (6 lb 11 oz) and the Barking specimen 3·40 kg (7 lb 8 oz). Another large fish of about this weight was caught in the Dartford Tunnel region in late 1971; it measured 700 mm (28 in.) overall.

CARP FAMILY

Roach, *Rutilus rutilus*, is a fish which is widespread throughout Europe and, being adaptable in its feeding habits and requirements for spawning, occupies a wide range of habitats from small ponds, lakes, and reservoirs to rivers. In rivers it will occupy all except the fastest-flowing stretches. Because it is one of the most popular anglers' fishes in Britain, it has been widely introduced outside its original range, and is continually being 'restocked' within its natural range. The Thames has on many occasions been stocked by roach from London reservoirs and lakes by the Thames Angling Preservation Society, but on only a small scale, with marked fishes from Kew Gardens lake, between 1939 and 1973. The population of roach in the tidal Thames was thus largely native to the river during the period in which the survey took place.

From the earliest sampling at the Thames-side power stations, it was clear that the roach was already well established in 1967 at both the Wandsworth Bridge area and at the Blackwall Tunnel area. Thus, in the

TABLE 5.1 *Numbers of roach recovered from power stations on the tidal Thames 1967–73*

	Wandsworth Bridge	*Blackwall Tunnel*	*Outfalls*	*Dartford Tunnel*
1967	67	35	5	1
1968	169	61	6	2
1969	39	17	1	—
1970	49	18	2	—
1971	64	34	—	—
1972	58	7	—	—
1973	58	2	—	—

last three months of 1967, 67 fish were caught at Fulham and 35 fish at the Blackwall Tunnel area. In the same period 5 roach were caught at Barking. Succeeding years showed much the same proportions as set out in Table 5.1.

It is important to emphasise that these figures should not be compared too rigorously year to year, as external factors such as a decline in 'fishing effort', biased them after the first two years of sampling. However, a general inference that can be validly drawn is that the roach was well established in the river at least as far downstream as the Blackwall Tunnel area in 1967 and that the population was maintained throughout these years. There is no reason to doubt that the roach is still abundant in these reaches of the tidal river when freshwater flow and tidal conditions allow.

Some interest attaches to the occurrence of this species at West Thurrock (Dartford Tunnel area), where isolated living specimens were caught in 1967 and 1968 (Wheeler, 1969; Huddart and Arthur, 1971), and in November 1974 and February 1977 (M. J. Andrews – personal communication). Normally, the roach is regarded as intolerant of any degree of salinity, but in this region salinity varies (at half-tide and under normal freshwater flows being about 37 per cent seawater). Two explanations might be advanced for these occurrences. First, all took place in the winter months (November to February), as did most occurrences of other freshwater fishes, e.g. pike, carp, bream, gudgeon, and perch, when freshwater flows would be high and salinity lowered and the main river marginally more suitable for them to survive. Second, the high freshwater flows of this period could have carried isolated specimens from some of the lower Thames tributaries, e.g. the Darent, into the main river. The occurrence of these fish, while interesting, is not of great significance to the ecology of the river.

The majority of roach caught at Fulham were relatively large (Table 5.2); in the 1968 sample, for example, only four fish were below 10 cm (4 in.) in length. This may be due to biased sampling, for large fish are seen and picked out more easily than small fish, although this tendency would have been partly compensated for because power station intakes catch more small fish than large ones, i.e. the weaker swimmers are unable to avoid the intake current. The figures are, however, interesting in that in 1968 both the mean and the maximum lengths were lower than in 1971 and 1973 (when they were broadly the same). This could be interpreted as evidence that young fish had recently colonised the Fulham area in 1968 whereas by 1971 the roach had attained a stable mean length, and colonisation of the river had moved downstream. This suggestion is borne out by the figures for the Blackwall Tunnel area, which lies 21 km (13 miles) further downstream. In both 1968 and 1971 the mean length and the minimum and maximum lengths were strikingly less than samples in the same years at Fulham. It is tempting to suggest that the roach population here comprised relatively recent immigrants from upstream.

TABLE 5.2 *Lengths of roach at two sampling sites on the tidal Thames. Lengths given are body lengths exclusive of tail (standard length); some samples represent only part of the year's catch*

Locality	*Year*	*No.*	*Range*	*Mean*
Fulham	1968	96	70–280 mm (2·8–11 in.)	170 mm (6·7 in.)
	1971	43	170–310 mm (6·7–12·2 in.)	215 mm (8·5 in.)
	1973	59	120–300 mm (4·7–11·8 in.)	206 mm (8·1 in.)
Blackwall	1968	61	50–240 mm (2·0–9·5 in.)	158 mm (6·2 in.)
Tunnel	1971	34	60–250 mm (2·4–9·9 in.)	155 mm (6·1 in.)

In many species the age of an individual fish can be estimated by 'reading' the growth rings which are formed on the scales. As the fish grows, each scale becomes larger by the addition of new tissue at the edge and on the surface. In summer, when temperatures are high and food is usually abundant, growth both of the fish and of the scale is fast; in winter growth is slow or even lacking. The result of this seasonal variation in metabolism is that the growth rings (circuli or ridges) are far apart during the fast growing season and close together when growth is slow. This means that the scale shows a change in density when examined under a microscope, and the formation of an annulus for each period of winter growth is observable. Spawning also results in a metabolic change in the fish's growth and spawning annuli are also detectable in certain circumstances. The value of 'scale reading' in the present case is that by determining the age of a sample of fish and relating the age to the length it is possible to make comparisons with the growth of members of the same species in other rivers.

Scales from some 280 roach from summer samples at the Wandsworth Bridge area were examined (and others from the Blackwall Tunnel area and Syon Reach). The mean standard length of each age-group is set out in Figure 5 and compared with age–length curves for other English rivers (from Mann, 1973). In common with other populations, the tidal Thames roach grew fast for the first four years of life, but thereafter their growth rate bears no comparison with those of other rivers, with the exception of those in the Thames at Reading. In the tidal Thames in the Wandsworth Bridge area, it seems that growth decreases sharply after the fourth year of life and continues by very small increments each year thereafter. This conclusion is derived from samples collected in 1972–3, and may not necessarily apply to the roach in this area subsequently. The figures for the much smaller samples from the Blackwall Tunnel area are also shown in figure 8. Here again the roach grows well early in life but

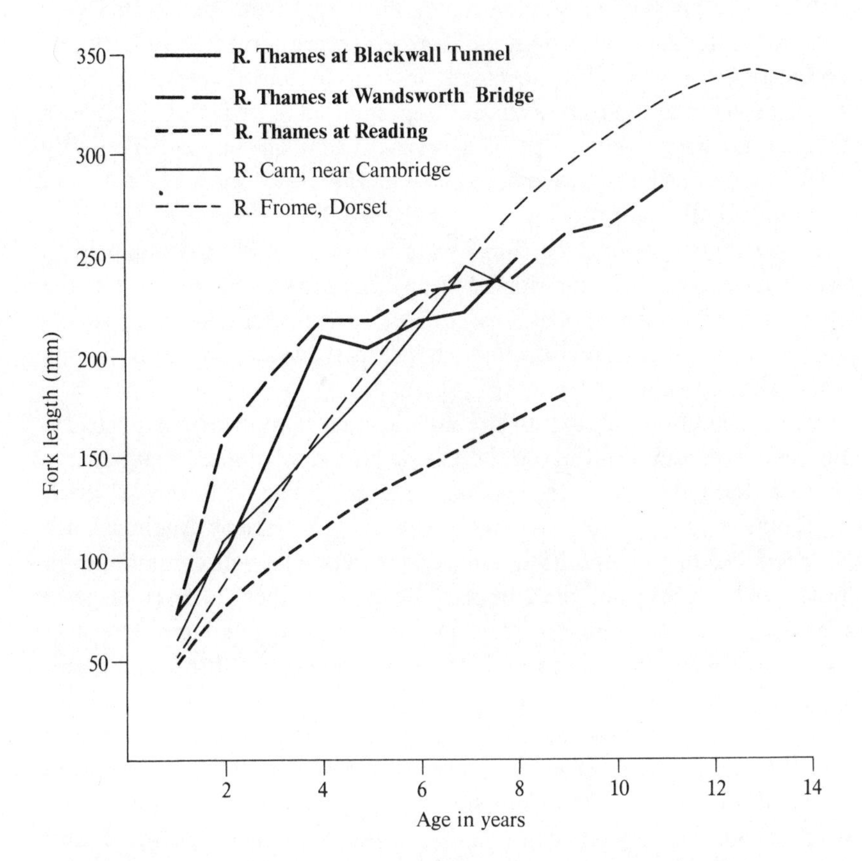

FIGURE 8 *Growth of roach in the tidal Thames (and other rivers: after Mann, 1976)*

growth is very limited beyond the age of four years. The length increment in each year is less than further upstream, which suggests both that the fish were living in the area, and had not individually swum downstream before capture, and also that conditions were marginally less suitable for the roach here than around Wandsworth Bridge.

The most plausible explanation for the poor growth of the roach in these parts of the tidal Thames is the food available. In undisturbed habitats young roach eat diatoms and planktonic crustaceans, graduating to aquatic insect larvae, molluscs, plants and algae, as they grow larger (Wheeler, 1969; Mann, 1973). The middle reaches of the tidal Thames contained abundant planktonic crustaceans (Cladocera, ostracods, and copepods) in surface tow-nettings in 1972 (Harris and Wheeler, 1974), and it seems certain that the growth of young roach would not be impeded by sparsity of suitable food. However, from their third year onwards suitable food would be scarce. Insect larvae were virtually absent in the area, molluscs occurred in clear-bottomed areas, but much of the river bed was covered in deep mud, and aquatic plants and filamentous algae were respectively non-existent and scarce. The only food resource which was abundant was the dense beds of tubificid worms, and although it is known that the roach did feed on these, there are reasons for thinking that this fish is not particularly well adapted to eating these bottom-living worms in any quantity. The reason for the characteristic growth of tidal Thames roach seems therefore to be associated with abundant suitable food in the early years, followed by inadequate food supply later in life.

These conditions applied to the middle reaches, but further upstream between Kew and Teddington there is no reason to suppose that scarcity of food limited growth to a similar extent. Indeed catches of roach exceeding 1 lb (453 gm) weight by the Francis Francis Angling Club between Teddington and Richmond Locks were relatively common, even in the 1950s. However, the summary of these catches (p. 134) suggests that the pollution of the river here in the 1950s had an adverse effect on the growth of roach, seen in both the average weight of fish caught, and in the number in excess of 1 lb (453 gm).

The tidal Thames above Richmond contains a large stock of roach, and anglers occasionally make above-average catches. Recent catches reported included 23·6 kg (52 lb) comprising 74 fish, an average weight of just over 311 gm (11 oz) per fish (*Angling Times*, 19 November 1975), and 92·4 kg (204 lb) of roach and dace caught by fifty competitors in a fishing match (*Angling Times*, 10 December 1975).

Bleak, *Alburnus alburnus*, is common throughout most of lowland England and also widely distributed in Europe. Typically it is most abundant in slow-flowing rivers in which it occupies the upper metre or so of water, although when the river is in spate it will retire to sheltered deeper water. Its food consists mainly of planktonic organisms, especially small crustaceans, insect larvae, and aerial insects with occasional terrestrial insects and other arthropods which fall into the water from overhanging trees, bankside vegetation, or as a result of high winds. The bleak is, however, an adaptable fish and Cragg-Hine (1969) showed that in an artificial canal at Peterborough this fish ate large quantities of aquatic plants, including filamentous algae, midge larvae, and on one occasion fish; during the angling season these bleak also fed heavily on maggots and bread from ground bait!

In the tidal Thames bleak were present in some numbers from 1967 onwards. The samples collected at the electricity generating stations were, however, relatively small, and it may be that being surface-living fish they tended to avoid being captured in the near-shore water intakes of the stations. The capture of bleak at these generating stations is set out in Table 5.3. Although these figures are small and probably do not represent a true sample of the abundance of the species in the metropolitan reaches, a number of points of interest can be observed. In the period 1967–9 the bleak was obviously not so common in the Blackwall Tunnel area as it was upstream at Wandsworth Bridge, but in the succeeding years more were caught further downstream. This suggests that after 1969 the river improved sufficiently for the bleak to move downstream in greater numbers. The condition of the river in the Outfalls area was, presumably, unsuited to this fish, for in contrast to the other freshwater fishes, e.g. roach and bream, no bleak was captured there.

TABLE 5.3 *Catches of bleak at power stations on the tidal Thames 1967–73*

	Wandsworth Bridge	*Blackwall Tunnel*	*Outfalls*
1967	10	—	—
1968	5	1	—
1969	9	2	—
1970	9	5	—
1971	12	14	—
1972	10	5	—
1973	8	2	—

This apparent scarcity of the bleak in the metropolitan reaches of the river was disproved by the number of fish caught by anglers during the fishing experiments organised by Mr Henri Jaume, then of the Press

Office of the Greater London Council, and the officers of the Thames Angling Preservation Society. From 1968 to 1975 each year teams of anglers fished the Thames in the heart of London solely to establish if fish were present. In 1968 in adverse weather and tidal conditions, no fish were caught; in each succeeding year (except for 1973 and 1975) bleak dominated the catches. The catches year-by-year for this species were: 1969 – 159 bleak; 1970 – 46 bleak; 1971 – 42 bleak; 1972 – 339 bleak; 1974 – 13 bleak. Their distribution in the river was of equal interest. All the fish in 1969 were captured between Wandsworth and Chelsea Bridges, that is upstream of central London. Those in 1970 were all caught in the vicinity of Chelsea Bridge, but in 1971 two were caught at Lambeth near the bridge, while others were taken from the Albert Embankment (3), near the Festival Gardens, Battersea (14), and at Putney Bridge (28). The most significant catch was those taken at Lambeth, well into central London, but this was eclipsed by the next year's (1972) catches when large numbers of bleak were caught at Battersea (47), Vauxhall Bridge (101), on the shore outside County Hall (144), at Waterloo Bridge (29), at Southwark Bridge (3), and on the foreshore beneath the Tower of London (28). These fish were the first to be caught in the heart of the City for probably more than a century. The 1974 catches were relatively disappointing, but fish were again caught at Lambeth Bridge.

In August 1977 a small fishing experiment was organised by the Thames Angling Preservation Society and again bleak dominated the catches; at the Albert Embankment (26), the County Hall (19), and the Festival Hall (21).

These fishing experiments showed that during the period 1969 to 1977 bleak were abundant in the river in the heart of London and, taken with the power station catches, showed that this adaptable little fish had successfully colonised the tidal river in its middle and upper reaches.

Other reports confirmed this; in September 1970, after heavy rain which had raised the river level, gulls were seen snatching bleak from the surface at Lambeth Fire Brigade Pier, where an angler caught two bleak, and several fish which could only have been bleak from the description given, were seen between Tower Pier and Customs Pier.

Upstream of London netting the river showed that bleak dominated the fish fauna at least numerically. The average catch for a haul of the seine-net in the Syon Reach – Mortlake Reach – Corney Reach area (between 16·9 and 22·5 km (10·5–14 miles) above London Bridge) was 98 fish, and the maximum catch exceeded 376 fish in one haul in Syon

Reach in July 1972. Above Richmond Lock netting showed that it was as abundant here, if not more so, the maximum number in a single haul being 403 fish (July 1972). More recently, a netting of the weir-pool at Teddington in September 1977 by the Thames Angling Preservation Society yielded 146 bleak (L. Harris – personal communication). In these reaches above London many bleak are caught, and were caught by anglers during the period 1967–7 (and earlier above Mortlake), but the bleak is a small fish of little interest to anglers and most such catches are unrecorded.

Estimation of the age related to the length of the fish by means of reading the annual checks on the scales showed that tidal Thames bleak grew rather faster and attained a larger size than bleak studied by Williams (1967) in the Thames at Reading. The length for age figures are set out in Figure 9, compared with the results from Reading. The explanation for this probably lies not so much in the better living conditions of the tidal Thames as in the excessively overpopulated river at

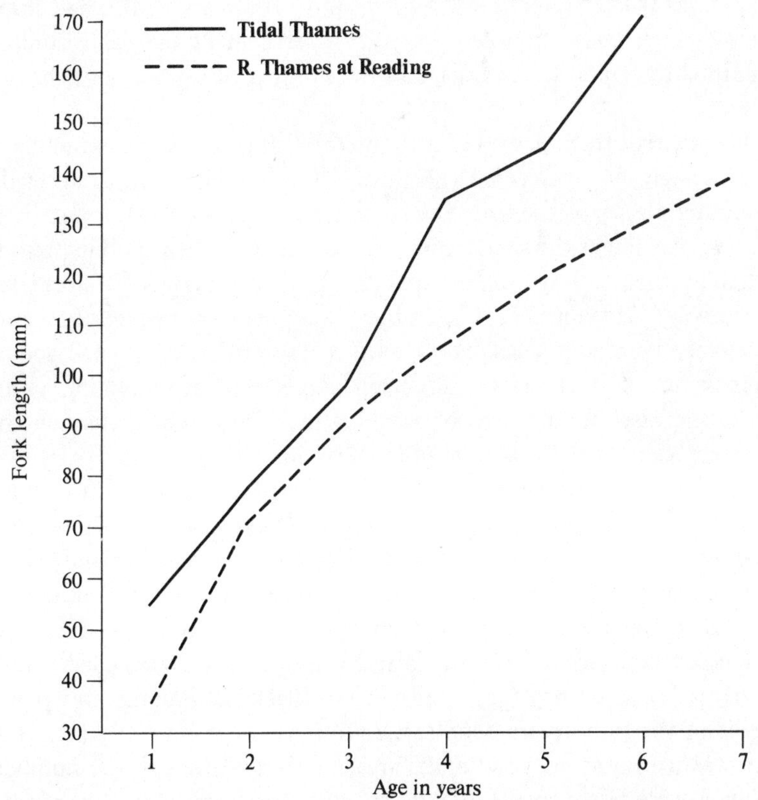

FIGURE 9 *Growth of bleak in the tidal Thames(and other rivers : after Williams, 1967)*

Reading, for Williams estimated the bleak population there to be around 7 fish per square metre.

Tidal Thames bleak were found to suffer from parasitic infestation by the larvae of the tapeworm *Ligula*. This parasitic worm finds its final host in the gut of a fish-eating bird, its eggs dropping into the water with the bird's faeces. The egg hatches out and in the first stage of its development may be eaten by a copepod crustacean, which, if eaten by a fish, will result in infestation of the fish. The parasite in its larval, fish-living stage is creamy coloured, flattened and may be up to 60 mm in length; it lives in the body cavity, not in the gut. *Ligula* is a very common parasite in members of the carp family in Britain and Europe, and has been found in the tidal Thames in roach as well. However, its discovery in bleak was of some interest in that it had not been reported in this species in Britain before. Young fish aged one or two years were worst affected and in samples from Petersham Meadows, Richmond, and Syon Reach between 47 and 84 per cent of the fish were infested with between 2 and 3 worms per fish. Older fish in their third, fourth, and fifth years of life had lower levels of infestation (between 45 and 70 per cent of the fish examined contained the larval tapeworm), and the number of worms per fish were fewer.

One effect of the tapeworm larvae on the fish is to result in swelling of the abdomen, but in severe cases the swelling is so gross as to upset the swimming ability of the fish. Such small fish were seen swimming in an ungainly fashion at the surface, having lost much of their mobility and in particular their ability to dive quickly (probably because of increased pressure on the swim-bladder). As they swam one saw frequent flashes of the silvery belly and sides, as opposed to the rare glimpses one gets of silver from a normal, active fish. Such impairment of swimming ability clearly increases the chances of a fish-eating bird catching the fish and thus perpetuating the life-cycle of the parasite.

The dace, *Leuciscus leuciscus*, is widely distributed across Europe and northern Asia almost to the Pacific coast. It typically populates clear, fast-flowing streams in large schools, but also inhabits lowland, slow-flowing rivers in numbers. In the tidal Thames it is relatively abundant, although it was not an early coloniser of the cleaner river. In the power station survey it occurred in the Wandsworth area in 1968 (4 fish), 1971 (2), 1972 (4), and 1973 (32), while in the Blackwall Tunnel area it was caught in the same years 1968 (2), 1971 (10), 1972 (1), 1973 (2). Only once was it caught in the Outfalls area, in 1972. Although the numbers involved were fairly small, they do suggest that from 1971 onwards the river had improved sufficiently as far downstream as the Blackwall Tunnel to support this rather sensitive species in some numbers.

Records of the capture of dace upstream of London are numerous. Netting in the Mortlake–Richmond area produced small numbers (5–10) in each haul of the net, which was certainly only a fraction of the dace population as these fast-moving, shy fish are difficult to catch in a slow-moving seine net in the current of the tidal river. Anglers in the Richmond Bridge area in September–October 1972 caught large numbers of dace in two fishing matches; on both occasions over 45·3 kg (100 lb) being caught by ten men (*Angler's Mail*, 4 October, 1 November 1972). In November 1972, one angler caught 80 dace weighing 8·593 kg (18 lb 15 oz 8 dr), this representing an average weight of 107·4 gm (3¾ oz), which, while not specimen fish, were nevertheless of reasonable size for the species in match fishing. Large catches of dace have been reported since in this region.

The Francis Francis Angling Club's records of dace in the Richmond to Teddington area (which are more fully discussed on page 134) showed that between 1953 and 1959–60 they caught relatively large numbers, with a gradual decline from then until 1970–1, although these records refer only to the 'best' fish of each season's fishing. However, the average weight of each fish in each season showed a small overall improvement between 1953–4 and 1964–5, with a decline until 1968–9. A similar decline at this period was observed in the average weight of roach.

Further downstream, in central London, dace were caught in the 'fishing experiments' of the Greater London Council, in 1971 (22 fish), 1972 (10), 1973 (1). In 1971 all 22 dace were caught on the foreshore at Putney in Wandsworth Reach; in 1972 they were caught off the Festival Gardens Pier, Battersea, and beside Vauxhall Bridge; the single fish in 1973 was caught from Chelsea. A similar experiment organised by the Thames Angling Preservation Society in August 1977 resulted in dace being caught from the foreshore near the Festival Hall, County Hall, and the Albert Embankment, in Kings Reach and Lambeth Reach.

The release of storm water from the sewers into the river following heavy rain in June 1973 resulted in a temporary pollution of the river following which at least 50 dace (besides other fishes) were picked up dead on the foreshore by Battersea Bridge; others were found in Chelsea Reach.

It seems clear from these figures that dace were present in numbers in the upper tidal river during the period 1950–64, although not so abundant as in the 1880s (Wheeler, 1958). A decline in the number of large fish, although not necessarily in the total numbers of the species, continued until 1969–70, by which time the species had become more common in the metropolitan reaches of the river as the chemical condition improved due to pollution control. From 1971 the dace has been moderately common above London Bridge, but occurs less frequently down as far as the Blackwall Tunnel area, and exceptionally at the Outfalls area.

Scales from 103 dace from the tidal river were 'read' for age; the resulting growth curve is shown in Figure 10, where corresponding curves for the rivers Frome and Stour in Dorset, and the Thames at Reading are

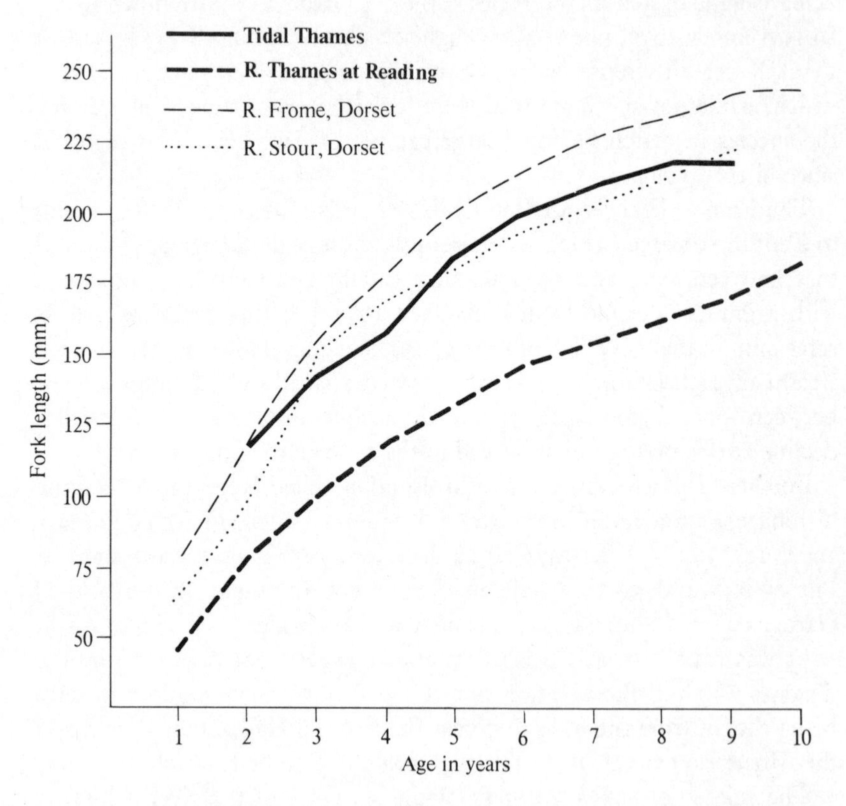

FIGURE 10 *Growth of dace in the tidal Thames (and other rivers: after Mann, 1974)*

shown for comparison. The growth in length with age in the tidal river is constantly better than in the non-tidal Thames at Reading, although there the limiting factor was almost certainly overcrowding, the density of dace being 0·21 fish per square metre in a total density of all species in excess of 10 fish per square metre (Williams, 1965). Growth of dace in the tideway was broadly comparable with that found in the elegant study of the species by R. H. K. Mann (1974) for the River Stour, Dorset, the only major discrepancy being in fish in their third year which were somewhat longer than the Stour fish. The diet of dace consists primarily of aquatic invertebrates, in winter trichopteran (caddis fly) larvae and molluscs, and in summer ephemeropteran (may fly) nymphs, blackfly and midge larvae (Mann, 1974). Such a fauna is to be found in the tidal Thames only in the uppermost reaches, although midge larvae and molluscs occurred in the Mortlake region in some numbers. The early good growth of Thames dace may be due to their feeding on the abundant zooplankton (copepod and cladoceran crustaceans, for example) in their first two years of life. What they feed on in later life is not known as the fish examined mostly had an empty gut, or had been eating maggots – presumably angler's ground bait.

The chub, *Leuciscus cephalus*, is a close relative of the dace and in general inhabits similar habitats, although large specimens especially are solitary rather than schooling fishes. In contrast to the relative abundance of the dace in the tideway, the chub proved to be rather uncommon. In the power station survey chub were caught in the Wandsworth Bridge area between November 1968 and January 1969, one fish 240 mm body length (9·5 in.), between October 1970 and April 1971, one fish 380 mm (15 in.), and on 17 March 1971, one fish 61 mm (2·4 in.). At the Blackwall Tunnel area one fish was taken in April 1972, 400 mm (15·75 in.) in length.

Other sources of information suggest that this species is more common upstream of London although never abundant in the tideway. Occasional specimens are caught by anglers; the Francis Francis Angling Club's records show specimens weighed-in in 1964–5 as one fish of 264 gm (9·3 oz), and in 1971–2 one fish of 455 gm (1 lb).

The bream or bronze bream, *Abramis brama*, is a widely distributed fish in the British Isles and across northern Europe outside the Arctic regions eastwards to the Caspian and Aral Seas. It is especially common in slow-flowing rivers and still waters, such as large lowland lakes and reservoirs. It is well adapted to feed on the river or lake bed, its upper jaw being protrusible so that the mouth forms a downward pointing tube with which it sucks in bottom-living insects, like midge larvae, small molluscs, and worms. The blood red tube-dwelling tubifex worm, which was so abundant in the freshwater reaches of the tidal river, was an important element of the food of the bream here, as it was of other fishes capable by the structure of their mouths of eating them. So abundant were these worms when the tideway was severely polluted that at low tide the exposed mud of the river bed looked bright-red with millions of worms exposed from their tubes. They were able to survive in polluted conditions because the tail end of the worm is protruded as a gill to absorb dissolved oxygen especially during low tide when the mud is exposed to the air. Most of the dissolved oxygen that these worms consume thus comes directly from the atmosphere, not from the river water as has been assumed in some quarters.

The bream proved to be common in the metropolitan reaches of the Thames from 1967 onwards, although the numbers caught in the power station survey continued to increase until 1972. The total catches at the main sampling areas are set out in Table 5.4. In addition, bream were caught at the Blackwall Tunnel area in December 1975 and May 1976; at the Outfalls area in January (2), February (1), and April 1975; and at the Dartford Tunnel area in March 1977 (M. Andrews – personal communication).

The figures from the electricity generating stations (which should not be analysed too rigorously on account of the methods of collection) suggest that in the Wandsworth Bridge area the bream was well established as early as 1967. Further downstream, at the Blackwall Tunnel area there was a steady increase in numbers between 1967 and 1971, presumably as more fish moved downstream to colonise the increasingly cleaner water. The isolated captures at the Outfalls area and at the Dartford Tunnel may have been due to Thames fish moving downstream, or they may have originated from the docks or lower tributaries. Certainly the two fish caught downstream at the Dartford Tunnel (1970 and 1977) could not be considered as normal members of the fauna there due to the increased salinity of these reaches.

TABLE 5.4 *Occurrence of bream at power stations on the tidal Thames 1967–73*

	Wandsworth Bridge	*Blackwall Tunnel*	*Outfalls*	*Dartford Tunnel*
1967	3	4	—	—
1968	13	7	1	—
1969	17	8	1	—
1970	16	9	—	1
1971	21	12	1	—
1972	13	1	—	—
1973	8	—	—	—

A few isolated captures of bream were made elsewhere in the tideway. A 457 mm (18 in.) long specimen was caught in summer 1971 at Bankside power station, Southwark (*South Eastern Power*, June 1971). Anglers' catches in the Richmond to Teddington region always included bream, and the Francis Francis Angling Club records show considerable numbers caught by the members of this specialised club from 1953–4 to the 1970s, the average weight of the largest fish having steadily increased through this period.

The bream caught in the metropolitan reaches of the tideway using power stations were mostly medium to large fish. Of the few small fish less than 150 mm (6 in.) in body length, most were caught in 1970 and 1972. Their occurrence year by year (all areas totalled) was as follows: 1967 (1), 1968 (2), 1969 (2), 1970 (10), 1971 (2), 1972 (11), 1973 (1). This suggests that after 1969 young bream were making their way downriver more readily, possibly on account of the cleaner water, or alternatively because the species was spawning further downstream. Unfortunately, few very young bream were caught either at the power stations or by other means. The tally was as follows; 38 mm (1·5 in.) netting in Mortlake Reach in June 1971, 42 mm (1·7 in.) in 1969, 2 at 50 mm (1·97 in.) in 1970 all at Fulham power station, and 51 mm (2·0 in.) in Chiswick Reach in May 1971, but although these must all have been O-group fish which had not yet had their first 'birthday' their capture tells us little about the spawning places of the bream in the river.

Of the larger fishes the greatest numbers were between 300 and 400 mm (11·8 and 15·8 in.) in body length, and 48 fish fell into this category. The largest bream caught in the tidal Thames (1971) was 410 mm (16·2 in.) body length and weighed 1·47 kg (3 lb 4 oz). An interesting capture was made in the autumn of 1968 of 5 large fish which together weighed 3·96 kg (8 lb 12 oz).

These figures suggest that the lower tideway was populated by large adult fish from prior to 1967 at the Wandsworth Bridge region and in increasing numbers at the Blackwall Tunnel from 1967. Presumably the limiting factor during this colonisation was the minimum level of dissolved oxygen in the water, although the presence of noxious or poisonous substances in the water could have also been involved.

The carp, *Cyprinus carpio*, is a native fish of eastern Europe and central Asia which has been widely redistributed by man first for food and latterly for angling. In Britain it is widely distributed, but is least common in Scotland and the western region, and numerous self-sustaining populations exist. Typically, the carp is an inhabitant of slow-flowing lowland rivers, oxbow lakes and backwaters, as well as lowland lakes in general, but so much redistribution of the species for angling purposes has taken place that it is found today in many atypical habitats. In the tidal Thames the carp proved to be relatively common, although its presence in any numbers had not been anticipated from earlier records.

The power station survey yielded 11 fish between 1968 and 1973, with a distribution along the river and in time, as follows; Wandsworth Bridge area, 1968 (1), 1970 (2), 1972 (1), 1973 (2); Blackwall Tunnel area, 1968 (2), 1971 (1); Outfalls area, 1968 (1); and Dartford Tunnel area, December 1969 (1). The number of fish caught is too small for further analysis, but there is some interest in the length distribution of these carp. Four were smaller than 127 mm (5 in.) in body length; the remainder were in excess of 230 mm (9 in.) with an average body length of 366 mm (14·4 in.), the largest being 485 mm (19·1 in.). The four smallest fish were caught between the Blackwall Tunnel area and the Dartford Tunnel area and may have been fish released into the lower tributaries or floodplain lakes by angling societies and which escaped into the Thames during high freshwater flows. All were caught between November and April when such freshwater flows are most common.

Much larger carp than these captured by power stations were present in the tideway. Marlborough (1972) reported that they were 'present in creeks by Fulham and Battersea Gas Stations'. On 16 June 1969 an angler caught 5 carp of which the largest was a mirror carp of 6·34 kg (14 lb) on the river front by the *Daily Telegraph* garage at Chelsea (*Angling Times*, 26 June 1969). A 2·04 kg (4 lb 8 oz) fish was netted out of the river at Millwall in June 1971 (*Angler's Mail*, 24 June 1971). In 1973 an angler fishing in the creek beside Lots Road power station caught 3 carp between 1·8 and 4·5 kg (4 and 10 lb), and as a result of pollution of the river after storm water had been discharged from the sewers following heavy rain a 477 mm (18·8 in.) body length carp was found dead at Lots Road on 22 June 1973. A 3·624 kg (8 lb) carp was picked up dead on the foreshore at Battersea the same day. In September 1976 a 6·4 kg (14 lb 2 oz) fish of 58 cm (22·8 in.) body length was foulhooked at Cheyne Walk, Chelsea by an angler who was fishing between the moored houseboats. Further upstream carp are relatively frequently caught between Richmond and Teddington. Examples may be given here; a 4·13 kg (9 lb 2 oz) carp below Richmond Lock (*Angler's Mail*, 13 September 1968); three members of the Francis Francis Angling Club caught carp of 1·099 kg (2 lb 6 oz 13 dr), 1·534 kg (3 lb 6 oz dr), and 2·313 kg (5 lb 1 oz 11 dr) at Twickenham (*Barnes and Mortlake Herald*,

30 July 1970); a 3·47 kg (7 lb 11 oz) common carp was caught at Radnor Gardens, Twickenham in March 1971 (*Richmond Herald*, 18 March 1971). The number of carp in the Twickenham area between June 1970 and March 1972 is borne out by the catches of the Francis Francis Angling Club of 11 fully scaled fish, the largest up to 4·757 kg (10 lb 8 oz), and 3 of the variety known as mirror carp, the largest up to 1·595 kg (3 lb 8 oz) (S. Callick – personal communication, 30 March 1972). The records of this angling club's catches starting in 1953–4 show that carp were rather uncommon fish in the Richmond to Teddington region between 1953 and 1968–9 but then increased in numbers remarkably. Despite allegations that these fish may have been introduced from Kew Gardens lake (which holds a large number of carp) following the Thames Angling Preservation Society's netting of the lake in 1969, there is no reason to believe this was the case as only about a dozen carp were transferred with the roach to the river.

The crucian carp, *Carassius carassius*, is by comparison with the carp a small fish. It inhabits still waters, especially small lakes, but presumably its natural habitat is oxbow lakes and marshy pools in the lower flood-plain of rivers, habitats which are excessively rare in an undisturbed state today. With the exception of some fenland drainage channels, the fish is not common in rivers. For this reason its occurrence in the tidal Thames is surprising, although relatively few specimens have been caught. Downstream, captures of crucian carp were made only at Ford's Works, Dagenham, in June 1973 (body length 223 mm), November 1974, and February 1975. The fact that only this power station captured this species suggests that they were fish which had moved down the River Beam which runs alongside the Ford Motor Company's site, and were part of a stocked population in the river or a nearby lake and not truly Thames fish. Crucian carp do, however, feature in anglers' catches in the upper reaches of the tideway on rare occasions, viz. between June 1970 and March 1972 2 were caught at Twickenham, the larger weighing 516 gm (1 lb 2 oz) (S. Callick – personal communication), and a single fish in netting the pool below Teddington Weir in September 1977 (L. Harris – personal communication).

Surprisingly, goldfish, *Carassius auratus*, which are certainly not native fish in the British Isles, proved to be more common than their native near-relative, the crucian carp. The species is found naturally in China, but has been widely introduced throughout the world and in temperate regions will breed and establish self-perpetuating populations. Most goldfish in Britain are imported from the Bologna region of northern

Italy, or the United States of America. Presumably the tidal Thames goldfish were pet fish that had been released from a goldfish bowl, although the possibility that stocked ornamental ponds had overflowed into the river or its tributaries also exists.

Goldfish were caught first in 1967 in the Blackwall Tunnel and Outfalls regions and continued to be caught through to 1971. The total of specimens taken in order downstream and by year were: Wandsworth Bridge, 1969 (3 fish), 1972 (2); Blackwall Tunnel, 1967 (2), 1968 (3), 1969 (1), 1970 (2), 1971 (3); Outfalls area, 1967 (1), 1970 (1), 1975 (1). The lengths of these fishes ranged from 71 to 139 mm (2·8–5·5 in.) with an average body length of 108 mm (4·26 in.). They all appeared to be plump and in good condition and with two exceptions were normal 'fish-shaped' fish, the exceptions having long and elaborated tail and other fins (the veiltail of the goldfish breeder). Several of these fish were yellow or golden-orange in colour, but others were greeny-brown.

As goldfish are relatively undemanding as regards temperature and dissolved oxygen levels it is probable that they were well suited for life in the tideway as it began to recover. It is an interesting speculation that once predators such as trout, pike, and perch became established in the river the goldfish's coloration proved a positive disadvantage; this might account for the apparent disappearance of the species after 1971.

The tench, *Tinca tinca*, is another member of the carp family which occurred rarely in the tidal river. Again, its presumed natural habitats are the oxbow lakes and flooded marshland pools of lowland floodplains, but a few live in the lower reaches of rivers. Its occurrence in the river is not therefore altogether surprising, but one would not expect it to be an abundant fish.

In the power station survey between 1967 and 1973 it was caught on seven occasions as follows; Wandsworth Bridge, 1967 (1), 1968 (1); Blackwall Tunnel, 1970 (1), 1971 (3); Outfalls area, 1972 (1). A single specimen was caught in the last-named region in November 1974 (M. Andrews – personal communication). Although these figures appear to show a gradual shift downstream between 1967 and 1972, it is clearly not safe to speculate on their significance as so few fish were involved. Records of the occurrence of the tench from anglers' captures and other sources are few and were confined to the upper reaches of the tideway, for example 'occasional between Twickenham and Teddington' (Wheeler, 1958), and 'a noted locality is Twickenham' (Marlborough, 1963). The Francis Francis Angling Club records show that their members catch tench on their monthly outings in this region. Catches of 'best fish' weighed-in follow: 1955–6 1 of 656 gm (1 lb 7 oz), 1960–1 1 of 798 gm (1 lb 12 oz), 1967–8 1 of 639 gm (1 lb 7 oz), 1970–1 1 of 864 gm (1 lb 14 oz), 1971–2 2 of 618 gm (1 lb 6 oz) and 913 gm (2 lb), which shows a general improvement in size of fish and suggests that the tench may have become more common over the period.

The gudgeon, *Gobio gobio*, is a small, bottom-living fish which is common in the middle reaches of rivers where current is moderate and the bottom alternately clean with exposed gravel beds and muddy in the bends of meanders. However, it is a versatile fish which can adapt to a variety of different riverine habitats and which will also live in lakes. In the tidal Thames it is clearly common in the upstream reaches but also ocurred in some numbers in the upper metropolitan regions. The power-station survey of 1967–73, however, suggested that intakes of most power stations were not well designed for catching such bottom-living fishes as gudgeon (and probably barbel and tench as well)! Power stations with the river water intake at the bankside did not catch many gudgeon, while one (Lombard Road, Battersea) that had its intake in the river bed caught considerable numbers. This survey yielded gudgeon at Wandsworth Bridge in 1969 (25 fish), 1970 (13), 1973 (1), Blackwall Tunnel area 1970 (1), the Outfalls area 1968 (1), and the Dartford Tunnel area 1970 (1). The occurrence of this fish on single occasions downstream of central London could be due to the displacement of a few fish from the upper river or from tributaries during high freshwater flows and, while interesting, is not significant. The gudgeon caught in the Wandsworth Bridge region were all relatively large fish, ranging in body length between 105 and 150 mm (4·1 and 5·9 in.), average 122 mm (4·8 in.).

Netting the tideway above London demonstrated how abundant the gudgeon was in the Syon Reach, Twickenham, and Richmond areas where anglers regularly catch it. Single hauls of the net in Syon Reach caught 46 fish, and at Petersham Meadows 109 gudgeon. Here the fish captured were a mixture of small and large fish. The Petersham Meadows sample contained 7 fish below 70 mm (2·8 in.), 64 fish between 70 and 100 mm (2·8–3·9 in.), and 38 fish between 100 and 150 mm (3·9–5·9 in.) in total length. The presence of so many young fish indicates that at least above Richmond Lock there was a stable, breeding population of gudgeon, and the same conclusion can be drawn for Syon and Mortlake Reaches, where a number of young gudgeon were also caught. This is in contrast to the presence of large fish only in the Wandsworth Bridge area, and suggests that here breeding did not take place.

The barbel, *Barbus barbus*, like the gudgeon, is a bottom-living fish, but unlike it is confined to rivers in regions where there is a moderate flow and generally clear gravelly bottom, and where water quality is at least moderately good. Its occurrence in the metropolitan reaches of the tideway was thus of some significance in demonstrating the improvement in quality of the water, although relatively few fish were captured. Isolated captures were made at Fulham power station (just downstream of Wandsworth Bridge) in November or December 1967, and between April and July 1968; both were moderately large fish, 425 mm (16·75 in.) and 440 mm (17·34 in.) respectively in body length. There is, however, reason to believe that the species was more common than those isolated captures suggest, and as already suggested, most small power stations are not very efficient tools for capturing bottom-living fishes. In June 1973, following a sudden pollution of the tidal river to the west of London by storm water from the sewers, dead or dying barbel were found at Lots Road, Chelsea, 370 mm (14·58 in.) and at Battersea, 906 and 566 gm (2 lb and 1 lb 4 oz), which showed that this species was living in the area.

Upstream, the barbel is only occasionally caught by anglers. One of the largest was 4·983 kg (11 lb) in weight was caught below Teddington Weir (*Angler's Mail*, 11 July 1973) and at about that period a number of small barbel about 152 mm (6 in.) long were caught at Richmond, with larger fish up to 1·529 kg (3 lb 6 oz), while other large fish were taken at Barnes Bridge and Isleworth. A number of small barbel were caught by netting the river between Richmond and Teddington, July 1972 – 2 fish 152 and 203 mm (6 and 8 in.), May 1973 – 1 fish *c.* 150 mm (6 in.), and September 1977 – 1 fish in the Teddington weir pool (L. Harris – personal communication).

The rudd, *Scardinius erythrophthalmus*, is a fish of still waters, especially marshy pools, which occur rarely in rivers except for some Fenland drains (which have ecological characteristics more in keeping with lakes than rivers). Its occurrence in the Thames was therefore unexpected, but it was captured only once in the power-station survey, at Lombard Road, Battersea (Wandsworth Bridge area) on 7 February 1972. This fish was 148 mm (5·8 in.) in body length. The reputed capture of a 906 gm (2 lb) rudd in the Thames below Tower Bridge by a ten-year-old boy (*Evening News*, 18 May 1971) seems to be beyond the bounds of probability in view of the scarcity of this fish in the tideway. The species was, however, known to inhabit the West India Docks in 1970.

The minnow, *Phoxinus phoxinus*, is a well-known inhabitant of rivers and some large, highland lakes, such as Lake Windermere. It is probably most abundant in small streams with a moderate-to-fast flow, and clean gravel bottoms over riffles, and well-oxygenated water. Its occurrence in the tidal Thames was therefore something of a surprise, although in keeping with its habitat requirements it was far from common. Netting in July 1972 in Syon Reach and above Isleworth Eyot yielded 3 and 4 minnows respectively and at Petersham Meadows, just below Teddington Weir 8 were caught in a single haul. Their body lengths ranged from 32–60 mm (1·3–2·4 in.) and all but one were smaller than 53 mm (2·0 in.). It is likely therefore that they were all relatively young fish. These captures show that the minnow was established in the upper reaches of the tideway in small numbers, but is not likely to spread much further downstream than Corney Reach in view of the physical character of the river below that region.

LOACH FAMILY

The stone loach, *Noemacheilus barbatulus*, has similar habitat requirements to the minnow, and in broad detail has the same distribution in the British Isles. The only occurrence of the species in the tidal Thames was of a single fish caught in Syon Reach, just below Isleworth Eyot in July 1972, its body length was 76 mm (3 in.). Possibly it is more common on gravel bottoms between Richmond Weir and Teddington, but it is unlikely to be found further downstream due to the nature of the river bed.

STICKLEBACK FAMILY

The nine-spined stickleback, *Pungitius pungitius*, is widely distributed in the northern hemisphere, especially in northern Europe and North America, while in Asia it occurs along the Arctic Ocean coastline (Wootton, 1976). In Britain its occurrence is widespread but distinctly erratic, but it occurs often in overgrown, densely weedy, muddy ponds and backwaters of rivers. In contrast to the wide range of habitats used by the three-spined stickleback, this species is relatively localised, although often common within any inhabited locality, but it does tolerate lower dissolved oxygen levels than its more abundant relative.

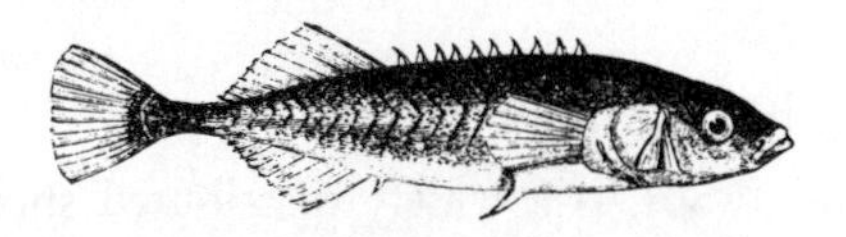

FIGURE 11 *Nine-spined stickleback*

In the tidal Thames the only specimens to have been captured have been taken at Ford's power station at Dagenham in January and October 1975 (each one fish), November 1975 (2), and March 1976 (3) (M. Andrews – personal communication). Their occurrence at this one locality is clearly due to the siting of the works beside the River Beam in which stream the species has been frequently captured above the tidal influence. The nine-spined stickleback is also locally abundant in the lower Ingrebourne, and in the Mardyke amongst Essex tributaries, and in the Cray at Crayford and the lower Darent, as well as in the drainage ditches and the 'tump' moats on the site of the old Woolwich Arsenal, now Thamesmead, on the Kent shore.

The three-spined stickleback is discussed in the next chapter.

BULLHEAD FAMILY

The bullhead or miller's thumb, *Cottus gobio*, is a small fish which usually reaches 100 mm (4 in.) in length, and in general is abundant in small rivers and large lakes with stony bottoms. It is often caught in the same habitat as the stone loach and tends to stay hidden under stones or in weedbeds during daylight, emerging in the twilight to forage for food. Its occurrence in the tidal Thames was therefore slightly surprising, and as it was captured further downstream and in some numbers it might be suggested that it is more adaptable than the stone loach, with which it is often bracketed because they live in similar habitats.

During the 1967–73 power station survey the bullhead was caught on five occasions all at Lombard Road, Battersea, just downstream of Wandsworth Bridge. In chronological order these captures were as follows: August–November 1969, 1 fish, 123 mm (4·8 in.) long, May–October 1970, 3 fish, 105, 115, 123 mm (4·1, 4·5, 4·8 in.), March 1971, 1 fish, 67 mm (2·6 in.). In addition, by netting in Mortlake Reach in May 1971 several other specimens were caught, 53, 60, 65 mm (2·1, 2·4, 2·6 in.) in length, and at Petersham Meadows, Richmond in September 1972, 51 bullheads of size range 26–55 mm (1–2·2 in.), mean body length 32·24 mm (1·27 in.) were caught. It is noteworthy that only one of this sample exceeded 40 mm (1·6 in.) in body length.

PERCH FAMILY

The perch, *Perca fluviatilis*, is a widely distributed species in northern Europe and northern Asia and is represented by a closely related species in North America. It is undemanding in its selection of habitats being found in lakes and rivers, although in the latter it is most abundant in lowland areas where the current is slow and bankside and near-shore vegetation offers cover. It seems relatively sensitive to water quality and requires a moderately high level of dissolved oxygen.

FIGURE 12 *Perch*

While the perch is not especially abundant in the tidal river, the power station survey showed that it occurs in small numbers as far downstream as the Outfalls area and becomes increasingly common upstream. The total catches from the regions in chronological order are set out in Table 5.5. Catches in the Outfalls area continued in the period 1974–7: November 1974 (1), January 1975 (3), February 1975 (1), April 1975 (1), January 1976 (1) (M. Andrews – personal communication). In addition perch were caught in the Dartford Tunnel area on 9 March 1970 (1 fish 202 mm (8 in.) body length), and in February 1977 (M. Andrews – personal communication). Even more surprising in view of the salinity of the water was the capture of a very healthy, actively swimming perch at Shell Haven, Canvey Island on 8 February 1978, a female 210 mm (8·3 in.) in body length with well developed roe.

In length the power-station-captured fish were relatively small for this species, which can attain 50 cm (19·7 in.) in length, but none were very young fish. The length distribution of these fish is set out in Table 5·6 which shows that there was no clear preponderance of any one length group other than in the 101–150 mm (4–5·9 in.) range of juvenile fish. This suggests that during the period studied there was a small but continuous recruitment to the perch population by migration from upstream. The larger numbers caught in the Blackwall Tunnel area in 1971 and 1972, and in the Wandsworth Bridge area in 1971, suggest that the improved

TABLE 5.5 *Catches of perch at power stations on the tidal Thames 1967–73*

	Wandsworth Bridge	Blackwall Tunnel	Outfalls
1967	4	—	1
1968	4	5	—
1969	1	2	—
1970	3	2	—
1971	8	8	—
1972	4	11	1
1973	2	2	—

chemical condition of the water had encouraged more fish to move downstream from the upper tideway.

In the upper tideway perch are caught by anglers, but judging from the sparsity of reports, neither very large fish nor very great numbers are captured. The Francis Francis Angling Club records of captures between Richmond and Teddington show that between 1953–4 and 1971–2 perch were caught in most years during their monthly competitions: 1953–4 (1); 1954–5 (4); 1956–7 (1); 1957–8 (1); 1958–9 (1); 1960–1 (2); 1961–2 (3); 1962–3 (2); 1963–4 (5); 1964–5 (4); 1965–6 (3); 1966–7 (1); 1967–8 (1); 1969–70 (1). The interesting trend of an increase in numbers from 1961–2 to 1965–6 is heightened by the increase in the number of

TABLE 5.6 *Numbers of perch in five length groups caught in the metropolitan Thames 1967–73*

Standard length (*mm*) (*in.*)	50–100 2–3·9	101–150 4–5·9	151–200 5·9–7·9	201–250 7·9–9·9	251–300 9·9–11·8
Wandsworth Bridge	7	4	6	6	3
Blackwall Tunnel	5	13	3	3	4
Outfalls	—	2	—	—	—

specimens caught which weighed in excess of 453 gm (1 lb); none until 1961–2 when 1 fish was caught; 1962–3 (2); 1963–4 (2); 1965–6 (1). The decline after 1965–6 parallels that seen in the roach, and may have been due to the effects of the severe winter of 1962–3, while the absence of perch after 1969–70 could have been as a result of an infection which affected the species on a large scale in southern England about that period.

The ruffe, *Gymnocephalus cernuus*, is a small relative of the perch which is found almost exclusively in rivers and canals and lives close to the bottom, feeding on bottom-living invertebrates. In the British Isles it is naturally confined to the eastern English counties, but in Europe it is widely distributed. During the power station survey of 1967–73 it was caught in small numbers in the earlier years in the metropolitan reaches but seemed to decline in numbers after 1970. The total catches were as follows; Wandsworth Bridge area, 1967 (2), 1968 (2), 1970 (2), 1972 (1), 1973 (1); Blackwall Tunnel area, 1967 (1), 1970 (1). A single ruffe was caught in the Outfalls area in November 1974 (M. Andrews – personal communication). The body lengths of these fish ranged from 44 to 111 mm (1·7–4·4 in.) and showed two clearly defined length groups, 44–62 mm (1·7–2·4 in.) 5 fish, and 101–11 mm (4·0–4·4 in.) 5 fish. The species does not appeal to anglers as its maximum length is *c.* 300 mm (11·75 in.). There is no supplementary record of its occurrence in the tidal Thames, although Mr J. Wade of the Francis Francis Angling Club tells me it is common in the Richmond to Teddington area but is not noted in their records when caught.

CARP FAMILY HYBRIDS

Hybrids between the bream and the roach, *Abramis brama* × *Rutilus rutilus*, occur relatively frequently in the tidal Thames. They were caught in the Blackwall Tunnel area, 2 fish (November 1967–January 1968), and in the Wandsworth Bridge area, 1 fish (December 1971–April 1972) and 1 fish (October–November 1972). They have been recorded in the Francis Francis Angling Club's monthly matches between Richmond and Teddington regularly in recent years: 1957–8 (2); 1958–9 (1); 1959–60 (7); 1960–1 (2); 1961–2 (5); 1962–3 (3); 1964–5 (5); 1965–6 (10); 1966–7 (5); 1967–8 (10); 1968–9 (6); 1969–70 (10); 1970–1 (9); and 1971–2 (5). Other specimens of this hybrid have been reported by anglers or sent in for identification mainly from the Richmond to Teddington area and it is clearly common in this region.

The less well-known hybrid between the bleak and the chub, *Alburnus alburnus* × *Leuciscus cephalus*, has been recognised from the non-tidal Thames at Walton-on-Thames, and from the River Mole (Wheeler, 1978b). In view of the general scarcity of the chub in the tidal river it is not expected to occur there.

THE FRANCIS FRANCIS ANGLING CLUB'S RECORDS

This angling club, which takes its name from the distinguished Thames angler Francis Francis (1822–86), fishes from club punts between Richmond and Teddington. A monthly match is fished during the open season and the individual captures weighed, witnessed, and the fish are identified by at least two of the club's officers. The records for the 'best fish' of each species between the period 1953 and 1978 have been made available to me by the officers and members of the Club through Mr John Wade, and this represents a valuable record of information collected over several years in comparable conditions. References have been made in the text to the numbers and size of the fish caught species by species, but the information on two species, roach and bream, is of such interest

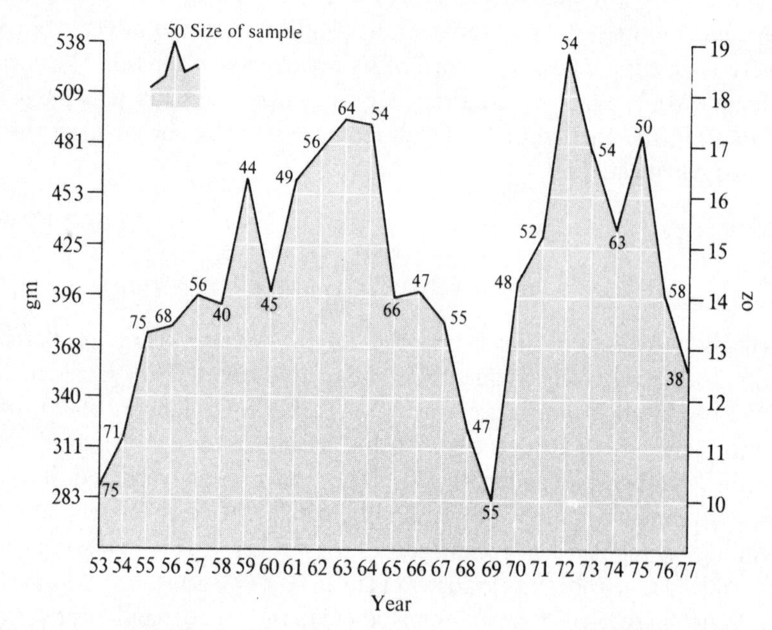

FIGURE 13 *Average weight of roach caught by the Francis Francis Angling Club 1953–77*

that it merits further discussion. The average weight of the roach and bream caught is shown in graphic form in Figures 13 and 14. For roach this shows a steady rise in average weight from 1953–4 to 1964–5, interrupted only in 1960–1, following which there was a dramatic decrease over five seasons until 1969–70, from whence a sharp improvement was noticeable

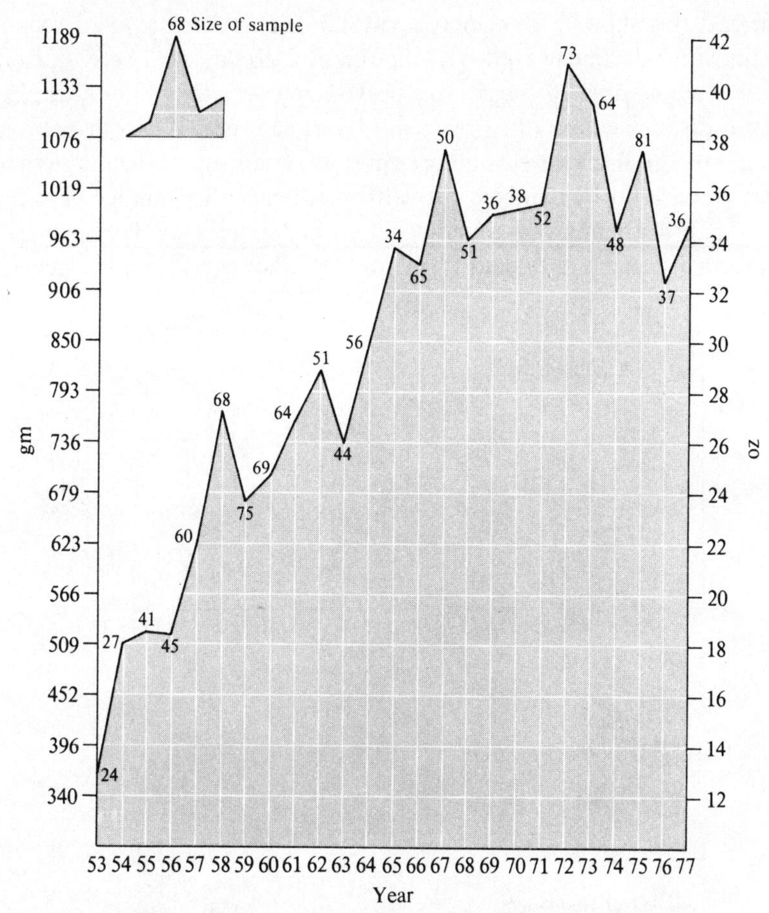

FIGURE 14 *Average weight of bream caught by the Francis Francis Angling Club 1953–77*

and was maintained until 1975–6 when it fell off rapidly. The bream, in contrast, shows no such dramatic decrease and the average weight of fish caught each season, with small decreases in 1959–60, 1963–4 and 1968–9, has shown a steady rise over a twenty-year period (1953–72) but has declined somewhat since. Expressed in simple terms the average weight of the 'best' bream in 1953–4 was 361 gm (12·8 oz), in 1963–4 it was 731 gm (25·82 oz), in 1972–3 it was 1,166 gm (41·184 oz), more than a three-fold increase in two decades, but subsequently by 1977–8 it had fallen slightly to 979 gm (34·6 oz). A similar picture results when the figures of the largest fish of both species are compared (Figure 15). Here the number of roach exceeding 453 gm (1 lb) in weight are plotted year by year and this shows a slow increase in numbers from 1953–4 to 1964–5 (again with a decline in 1960–1), followed by a decline until 1968–9, from whence the number rose again to a peak in 1972–3, after which there has been a distinct falling off. Bream, however, showed a sharp increase in the number of fish in excess of 453 gm (1 lb) from 1953–4 to 1957–8 and some fluctuation in numbers thereafter, although the number only fell below half of the maximum of 1975–6 during six seasons. Larger bream in excess of 906 gm (2 lb) and 1,359 gm (3 lb), however, showed a general

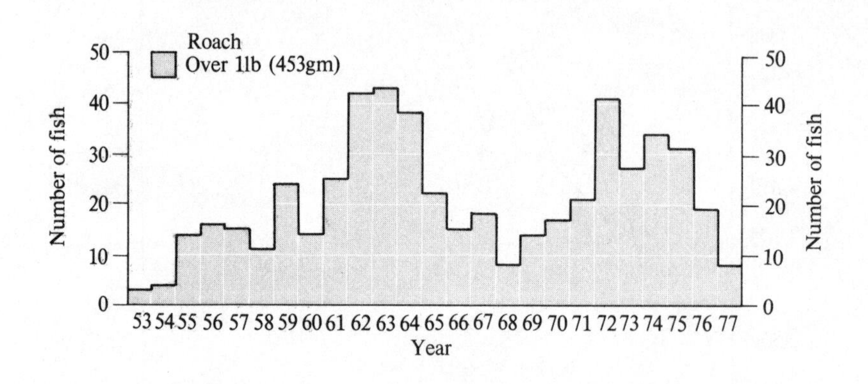

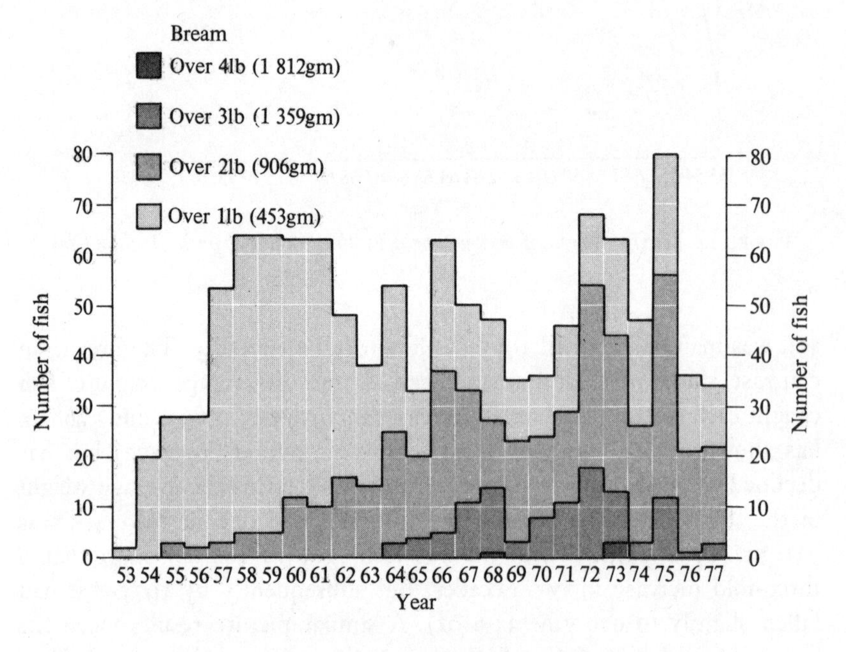

FIGURE 15 *Numbers of large roach and bream caught by the Francis Francis Angling Club 1953–77*

tendency to increase in numbers from 1955–6 and 1964–5 respectively but both declined sharply after the dry summer of 1976.

The explanation for these fascinating increases and fluctuations in the condition and number of both bream and roach is uncertain. Clearly, the increase in individual average size and in the number of roach exceeding 453 gm (1 lb) could have been due to the general decrease in pollution of the tidal river. However, this does not fully explain the change, because the river between Richmond Lock and Teddington Weir was not severely polluted (by comparison with the metropolitan reaches) and the improvement was noted as early as 1957–8, when the basic work necessary for the control of Thames pollution had not begun. A possible explanation for the major fluctuations in roach population might be that it had been affected by adverse weather. The exceptionally severe winters of 1946–7 and 1962–3 were both followed seven years later by an increase in the number of roach exceeding 453 gm (1 lb). It could be that the severe weather in those years affected the roach, so that there was a decline in the number and average size of the 'best' fish, observable within two years and continuing (as age-linked mortality) for a further four or five years, before fish hatched after the cold winters attained a sufficient size to appear in these figures. Attractive as this theory may be it does not seem to apply to either parameter in the bream, except that there was a slight but not sustained decline in the number of bream in excess of 453 gm (1 lb) for four years from the cold winter of 1962–3.

Since the 1975–6 season the average weight of roach and the number in excess of 453 gm (1 lb) has shown a dramatic decrease. A similar decrease can be seen in the numbers of large bream during the same two fishing seasons (1976–7 and 1977–8). There seems to be little doubt that this decline in the quality of these two species is due to the dry weather associated with the 1975–6 period which reduced freshwater flow over Teddington Weir and which produced a salt water incursion up to within a kilometer or two of the weir. This decline is most dramatic and draws attention to the fragile nature of the ecosystem in the upper tideway. It may be several years before anglers can catch large numbers of big roach and bream again.

The 'best fish' of each species weighed in through the season are itemised in Table 5.7. It represents the general proportions of the species in the upper tideway and illustrates the fluctuations in numbers, notably the decrease of large dace after 1964–5 (possibly as a result of the 1962–3 winter), and the increase in the numbers of large bream × roach hybrids until 1973–4, after which they became scarce. The numbers of perch, carp and tench (as well as chub and barbel which are not included in the table) are too small to allow any general conclusion to be drawn. The Francis Francis Angling Club's records form a valuable source of information on the fish in the upper tideway, objective information on which would otherwise be lacking.

TABLE 5.7 *Number of 'best fish' of each species in the upper tideway from the Francis Francis Angling Club's records 1953–4 to 1977–8*

Season commencing	*Species* *Roach*	*Bream*	*Bream ×Roach*	*Dace*	*Gudgeon*	*Perch*	*Carp*	*Tench*
1953	75	24	—	33	18	1	—	—
1954	71	27	—	34	14	4	2	—
1955	75	41	—	36	10	—	1	1
1956	68	45	—	33	13	1	2	—
1957	56	60	2	27	6	1	—	—
1958	40	68	1	23	6	1	—	—
1959	44	75	7	33	1	—	—	—
1960	45	69	2	18	4	—	1	1
1961	49	64	5	21	5	3	1	—
1962	56	51	3	21	3	2	—	—
1963	62	44	—	14	7	5	—	—
1964	54	56	5	12	10	4	—	—
1965	66	34	10	7	2	3	—	—
1966	47	65	5	5	4	1	—	—
1967	55	50	10	4	8	1	—	1
1968	47	51	6	3	13	—	—	—
1969	55	36	10	6	23	1	4	—
1970	48	38	9	16	16	—	8	1
1971	52	52	5	11	8	—	2	2
1972	54	73	9	34	26	—	—	6
1973	54	64	12	21	11	—	—	11
1974	63	48	4	32	27	1	5	8
1975	50	81	7	21	23	4	2	3
1976	58	37	6	26	11	3	1	3
1977	36	36	4	18	6	1	—	2

SIX

MIGRATORY AND ESTUARINE FISHES

The fishes included here comprise those species which migrate from the sea to breed in freshwater (such as the shads, salmon, and smelt), the eel, which breeds in the sea having spent a number of years in freshwater, and a group of fishes, such as the three-spined stickleback, the grey mullets, and the flounder, which are commonly found in estuaries as well as in the sea and freshwater. In every way these species are the most significant of all the fishes in the tideway, because they now inhabit or in their migrations pass through, the region of the river which was so grossly polluted in the 1940s and 1950s as to prove an impassable barrier to fish life.

JAWLESS FISHES

The lampern, *Lampetra fluviatilis*, is a fish-like vertebrate which lacks jaws, having a sucker disc on the underside of the head, well armed with teeth, with which it attaches itself to bony fishes, sucking their blood through the wound caused by the sharp central teeth. It is of wide distribution in northern Europe, but because it breeds in freshwater and makes a feeding migration to the sea it is now relatively rare in rivers where pollution and navigation weirs have prevented free access to the sea. In the Thames, from a position of great abundance, it is now a rare fish.

With one exception, all the lamperns found in the tideway have been taken in the Dartford Tunnel area. The period of their occurrence and the total length of the fish are as follows: November 1964, 1 of 290 mm (11·4 in.), November–December 1967, 1 of 283 mm (11·2 in.), May–October 1970, 1 of 250 mm (9·9 in.), and January–February 1973, 1 of 254 mm (10 in.). One was caught in January 1975 (M. Andrews – personal communication). In addition, Huddart and Arthur (1971) claimed that lamperns 'were found in some numbers' from November 1967 to January 1968, although they gave no figures to support this important statement. Indeed the only documented report they cite, other than those cited above from my own collections in the area, was of a 250 mm (9·9 in.) lampern caught in February 1970. One specimen was caught in the Outfalls area in February 1973, but was returned to the river alive.

It is interesting that most of these specimens were caught in the period November to February, and internal examination showed that they were in spawning condition. Although the numbers of fish are small and they are irregular in their occurrence, there seems little doubt that they represent an incipient breeding stock in the river. Hardisty and Huggins (1975) have found that in the mouth of the River Severn the main migration of lamperns takes place between mid-November and mid-February, December and January being the two months when most are captured. The onset of the main migration occurs after the first heavy rainfall of the autumn with consequent rise in river level. Thames lamperns thus appear to follow the same migratory pattern as the more abundant lamperns of the Severn.

EEL FAMILY

The eel, *Anguilla anguilla*, is an abundant fish in Europe, although most widely spread in the countries bordering the Atlantic Ocean, Baltic, North, and Mediterranean seas. It breeds in the Atlantic in an area south and east of Bermuda, and the postlarvae make the Atlantic crossing near the surface carried largely by oceanic currents. The postlarvae change into transparent elvers in coastal waters and these migrate into rivers during late winter and early spring in the British Isles. Although the eel is well known as a freshwater species, substantial numbers live in estuaries and on the sea coast.

Even during the period when pollution was at its worst in the Thames eels were found upriver. This suggested that some migration by elvers from the sea had been possible, but it is now known that substantial numbers of elvers had been imported into the Thames and the eels found in the 1940s and 1950s could have originated in this way. The eel proved to be present in variable numbers in the late 1960s all along the tideway where power station sampling was attempted. Upstream, catches in the Wandsworth Bridge area were as follows: 1967 (5), 1968 (8), 1969 (5), 1970 (8), 1971 (15), 1972 (9), 1973 (18). Downstream at the Blackwall Tunnel area catches were: 1967 (2), 1968 (5), 1969 (1) 1970 (3), 1971 (19), 1972 (13), 1973 (3); while in the Outfalls area they were: 1968 (12), 1972 (11), 1973 (4). In the latter area eels have continued to be captured regularly in numbers (M. Andrews – personal communication). Towards the mouth of the river, in the Dartford Tunnel area, eels were caught in 1967 (1), 1968 (1), 1969 (2), 1970 (1), 1972 (10), 1973 (1), and often in large numbers in 1974–7 (M. Andrews – personal communication), when on occasions over 50 eels were caught in a single day. During the power station survey of 1967–73 the numbers and sizes of eels preserved for examination were often small, for larger specimens were much appreciated for the table, so the above figures are minimal. There is also reason to believe that power stations' screens are not ideal for capturing the largest eels, because at both Barking and Fulham power stations, when the screens were dismantled for maintenance, numbers of very large eels were found in the bottom of the shaft and had probably been living there for some time. Mr A. E. Hodges told me that at Barking over 50 eels about 1,219 mm (48 in.) in length were found in these circumstances on 21 January 1971, and 6 eels 762 mm (30 in.) in length on 21 April 1970.

Large eels were captured on a number of occasions. For example, the sample of 7 fish from the Blackwall Tunnel area in April 1972 ranged in length from 691 to 1,040 mm (27·2–41 in.), with an average length of 838 mm (33 in.). A comparable sample of 9 eels at Wandsworth Bridge in March–April 1972 ranged from 605 mm to 990 mm (23·8–39 in.), average length 844 mm (33·2 in.). Records of large eels in the tideway are also available from angling sources, a 3·171 kg (7 lb), 1,067 mm (42 in.) eel stranded alive at Badcock's Wharf, Greenwich (*Angling Times*, 25 May 1967); a 1·869 kg (4 lb 2 oz) fish caught by angling at Battersea Bridge (*Angler's Mail*, 31 July 1969), and at Teddington 1·812 kg and 963 gm (4 lb and 2 lb 12 oz) by anglers (*Angling Times*, 18 May 1972).

The eel quickly spread into the lower tributaries of the Thames once it was again abundant in the tideway. In the mouth of the River Lee a 305 mm (12 in.) eel was caught in March 1969, and small specimens were caught in numbers by anglers in 1976–7 at Tottenham. On the other hand, they are said to have been found regularly in small numbers in both the tidal and non-tidal River Lee from 1964 to 1969 (B. Meadows –

personal communication). The River Roding was found to contain numerous eels as far up as Ongar, Essex, in the summer of 1976 (A. Dearsly – personal communication), and many elvers were caught with a dip-net in the River Mardyke, above Purfleet, in June 1973 and July 1975.

The capture of these elvers is really more significant than the presence of large fish in the Thames, for on account of the breeding habits of the eel they must have migrated upriver from the sea. The first elvers were caught at West Thurrock generating station in April and May 1968 (Huddart and Arthur, 1971). They were thought to have been dead at the time of capture, but they clearly had been living in the mouth of the river. Their length range was 64–70 mm (2·5–2·8 in.). In April–May 1972 many live elvers were found on the foreshore between Hammersmith Bridge and Richmond (D. Solomon – personal communication, and Solomon, 1976). Six elvers 6·4–7·6 mm (2·5–3 in.) in length were caught at Ford's Works, Dagenham (Outfalls area) in May–June 1972. Elvers were captured in the Outfalls area in April 1975, 4 fish length 66 mm (2·6 in.), in May (3) and June 1975 (2) and in March 1976 (1) (M. Andrews – personal communication). In May 1975 I received a verbal report of 'hundreds' of elvers swimming along the piers beside Woolwich Power Station, and in May 1977 again 'hundreds' of elvers were seen swimming along the edge of the bank of the lower River Wandle at Wandsworth (J. Chambers – personal communication). According to Andrews (1977), 'several thousand elvers were observed in the river both in 1975 and 1976 between London Bridge and Petersham', and he also recorded their capture in the effluent channels at Surbiton sewage works (these elvers having negotiated Teddington Weir) and at Beckton sewage works. It is of importance to point out in the context of visually noted elver migrations that the great peak of activity is usually at night. These records and reports show that there is now a substantial upstream migration of elvers in the Thames starting in April and possibly reaching a peak in May. This timing is in agreement with the April elver immigration into the River Severn, but, as Deelder (1970) has pointed out for Dutch elvers, the time of entry into rivers is strongly influenced by the water temperature of the coastal sea, being delayed after cold winters. Elvers in the Thames may therefore in future enter the river any time between February and April.

THE SHADS

The twaite shad, *Alosa fallax*, is a moderately large fish of wide distribution in European seas. It ascends rivers to spawn in the lower reaches in freshwater still within tidal influence in May or June. Like other migratory species it is now much rarer than it was in the nineteenth century, and in the tidal Thames it has declined from great abundance to comparative rarity. During the 1967–73 power station survey it was caught on three occasions at West Thurrock (Dartford Tunnel area), April–July 1968, November–December 1969, and May–August 1973.

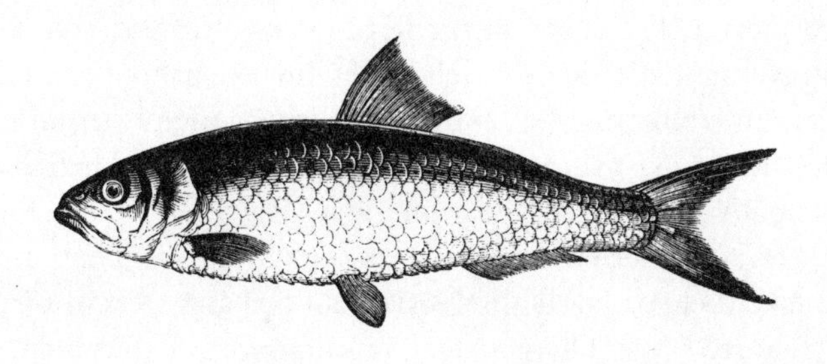

FIGURE 16 *Twaite shad*

These were young fish of body lengths 192, 212, 312 mm (7·6 8·4, 12·3 in.) respectively. A specimen of 309 mm (12·2 in.) was caught by an angler at Gravesend in October 1973. Since then Mr M. Andrews (personal communication) has informed me of captures at West Thurrock in October (2 fish) and November 1975 (1), and at Blackwall Point (Blackwall Tunnel area) in January 1976. It is clear that there is a strong tendency for the twaite shad to occur in the winter months in the mouth of the Thames (and occasional specimens are caught by anglers and others at Southend-on-Sea, and in Leigh Creek, at this time of year) but there does not as yet seem to be evidence of a spawning migration.

The allis shad, *Alosa alosa*, is larger, attaining a length of 600 mm (23·6 in.) and much rarer than the twaite shad. It has similar spawning habits but breeds far up in the freshwater region of rivers not within tidal influence. Probably because of the polluted state of many major European rivers (it is found from Norway to the Mediterranean) and the numerous navigation locks in their lower reaches, it is today very rare. In the tidal Thames, it has been reported on three occasions only; at West Thurrock in September 1975 and July 1976, and at Blackwall Point in February 1976 (M. Andrews – personal communication). It has also been captured at Kingsnorth power station in the lower Medway estuary in May 1975 (van den Broek – personal communication). Clearly, these captures are so widespread in season that they are probably no more than isolated captures of vagrant allis shad.

THE SALMON FAMILY

Without question the salmon, *Salmo salar*, is the most newsworthy of all British fishes. Its range encompasses the whole of the sea coast of northern Europe, from northern Spain northwards, Iceland, and the Atlantic coast of North America. It spawns in the headwaters of rivers, laying its eggs in small gravel nests ('redds') which are hollowed out by the adults. The eggs are laid in winter and hatch in early spring; after a period of one to three years the young fish descend the river to the sea. Their life in the sea is essentially a feeding stage which may last from one to four years before they return to the river of their hatching to spawn. Their migrations in the sea take them as far as the Norwegian Sea and west of Greenland.

As we have seen, the Thames stock of salmon was exterminated in the nineteenth century and the relatively few fish that were caught in the mouth of the river in the first decades of the present century were certainly wanderers from other rivers passing through the outer estuary. This was also the case of the salmon which was caught on 12 November 1974 on the screens at West Thurrock power station (Dartford Tunnel area). This fish was alive when found on the screens, and its gill covers were still moving when it was taken into the station chemical laboratory; its freshness was beyond doubt when I examined it later that day for it bled freely when an incision was made in the gill arches, which were also bright red (not dull as in a long dead fish). These points are necessary to make in view of later allegations in the press that this was the body of dead fish held for several months in a domestic deep freezer, although it is worth pointing out that the individual making this claim never saw the West Thurrock fish but 'recognised' it from newspaper photographs!

This Thames salmon was 787 mm (31 in.) in length and weighed 3·964 kg (8 lb 12 oz), which was, if anything, rather light for this length. Examination of its scales showed that it had spent two years in freshwater and two years in the sea. It was a female fish with well developed eggs in its body.

Its origin is unknown but it was probably one of the small stock of salmon which circulate on the English coast of the southern North Sea, and which are from time to time captured in coastal waters on the Norfolk, Suffolk, and Essex coasts. There is a contemporary record off Walton-on-the-Naze of a 2·265 kg (5 lb) salmon (caught with sea trout of 2·265 and 0·680 kg (5 lb and 1 lb 8 oz)) in October 1975 (*Angler's Mail*, 22 October 1975), but most of these salmon catches, for various reasons, go unpublicised.

Another salmon found in the tidal river dead on the foreshore at Dagenham in July 1975 was in a state of advanced decomposition. It measured about 533 mm (21 in.), but was so decomposed that no accurate weight could be recorded. However, there is no doubt that it was a salmon. Despite reputed statements by the Thames Water Authority (*Angling Times*, 22 October 1975) that 'tests prove it had lived in the river and had not been thrown in as a joke', no known test could have established this, although it is inherently improbable that anyone would throw a salmon away deliberately in view of the value of this fish! It must be concluded that the probability is that this fish had swum up the Thames from the sea before it died, but this cannot be proved.

A third salmon was found, also dead, in the Thames, this time above the tidal limit at Teddington Weir. This was found on the shore at the mouth of the River Ember which is a tributary of the River Mole, their confluence being within 366 m (400 yards) of the Thames where the Mole joins the Thames at Hampton Court. The fish was found by two anglers, Mr B. Swallow and Mr C. W. J. White, on 30 December 1976, and it was brought to me by Mr Ted Andrews, who realised the significance of finding a salmon there. It was a male fish which weighed 2·130 kg (4 lb 11 oz) and measured 600 mm (23·6 in.) in body length, 685 mm (27 in.) total length. When found it was fresh but the eyes were missing and there were wounds on the body, which could have been inflicted by scavenging birds. Again, its condition, that is, weight-to-length relationship, was poor. The gonads were moderately large and the fish had not spawned.

While it is not possible to prove that this fish had swum up the Thames to the River Mole where it was found, there are reasons for thinking that it was a 'Thames fish'. First, a mature cock fish such as this is unlikely to have been legally fished from a salmon river, nor is it likely to have been sold openly, for in the spawning season angling is not allowed, and the sale of breeding fish is illegal. Fish in this condition are thus difficult to obtain except by unlawful means. Even if the fish was unlawfully obtained it is unlikely that having gone to that much trouble anyone would simply discard it into a nearby river. It seems improbable therefore that this fish was disposed of by throwing it into the river, or was placed there as a hoax. On a more positive approach, the condition and position of the fish can easily be reconciled with its being a migrant upriver. A salmon, making a spawning migration upriver, would easily negotiate Richmond Lock, could jump over Teddington Weir or pass through the navigation lock, but would be confronted by Molesey Weir (which is virtually impassable to salmon). In search of a suitable spawning ground it would therefore turn into a tributary (in this case the River Mole and then the Ember) where it eventually died. The unspawned gonads support this interpretation, as does the find in late December, and the emaciated condition of the fish.

The River Ember fish thus probably had the distinction of being the first salmon to penetrate through London and into the non-tidal river for 140 years.

A fourth salmon seen leaping below Shepperton Weir by experienced observers in the early summer of 1978 was estimated to be about 4·5 kg (10 lb) in weight. Despite the efforts of Thames Water Authority staff to catch it, it remained at liberty. Presumably it was this fish which was found dead and partly decomposed on 5 September 1978 in the weir-pool at Shepperton. Dr John Banks, Fisheries Inspector of the Thames Water Authority estimated its length was 76 cm (30 in.) and its weight to have been in excess of 3·95 kg (8 lb 11 oz). It was four or five years of age and had spent two years in freshwater before migrating to the sea. The occurrence of this fish as far upstream as Shepperton is conclusive proof of the high quality of the water both in the tidal and lower non-tidal Thames.

The discovery of these four Thames salmon in the period 1974–6 is paralleled by occurrences in the River Medway. Here, in May 1973 a 12·684 kg (28 lb) salmon was found dead in the tidal reaches at Snodland, and a live specimen, weighing nearly 2·718 kg (6 lb) was captured by Kent River Authority staff also near Snodland. Taken overall these fish represent the occasional vagrants from the salmon 'stock' in the North Sea which have entered these rivers. Comparable numbers might occur in similar periods in the future, but their occurrence does not make them Thames salmon in the true sense of the term. The significance of their capture is that this most demanding of fish can now live in the tidal Thames and Medway, where they certainly could not twenty years ago. Only when it is possible for the salmon to reach the headwaters of the Thames to spawn and the smolts can migrate unaided to the sea will it be permissible to use the term Thames salmon again.

It is necessary in this context to record that young salmon have recently been released into the Thames. One thousand alevins were liberated at Richmond, and upstream more were released in River Eye in Gloucestershire, and in the Thames at Pangbourne and Tillingbourne (*Angling Times*, 23 November 1977). Two smolts presumably from one or more of these introductions were caught in the tideway in 1978. One was captured on 18 April at Brunswick Wharf power station in the Blackwall Tunnel area. It was 120 mm (4·75 in.) in body length and although the dusky blotches, known as parr marks, could still be seen, the fish was silvery overall and this suggested that it was approaching the life-stage called smolt, when it would migrate to the sea. Despite the quoted report of the Thames Water Authority (*Angling Times*, 7 June 1978) that the fish was swept up in a net and was kept alive for examination this was not so; it was captured on the screens at Brunswick Wharf generating station and was preserved in the laboratory there (I am indebted to Mr K. S. Gill of Brunswick Wharf for sending me the fish for examination). Another smolting parr was captured at West Thurrock generating station about the same time; this was about 150 mm (6 in.) in total length.

The trout, *Salmo trutta*, is as well known as its close relative, the salmon, and like it is widely distributed in northern Europe from the Arctic to North Africa and from Iceland to the Caspian Sea. The trout spawns in winter, usually later than the salmon and often further up-stream, but all kinds of variation in spawning place are found due to the wide variety of habitats in which the species occurs. *Salmo trutta* exists in two main forms, a migratory form, the sea trout, and a non-migratory form, the brown trout. An intermediate form is found in large lakes, often called the 'ferox' trout, which grows to a large size, is similar to the sea trout, and tends to substitute the lake for the sea, migrating into tributary rivers to spawn. These forms of trout are not genetically or morphologically distinct and are not now recognised as subspecies. Such differences as can be discerned between them are due to the environmental conditions in which they live.

In the tidal Thames numerous brown trout have been caught or reliably reported, and rather fewer sea trout. Reports of sea trout are always suspect because the large Thames trout are silvery in colour and superficially resemble sea trout, and only examination of the scales or the presence of marine parasites can prove that a true sea-going fish is involved. During the 1967–73 power station survey, trout were captured in the Wandsworth Bridge area: on 23 May 1970 1 of 143 mm (5·6 in.) body length, between July and December 1971 1 of 255 mm (10·1 in.), and 17 January 1972 1 of 430 mm (17 in.); at the Blackwall Tunnel area: on 15 May 1968 1 of 215 mm (8·5 in.), between April and July 1971 1 of 165 mm (6·5 in.), and between April and May 1972 1 of 155 mm (6·1 in.);

and in the Dartford Tunnel area between July and December 1971 1 of 445 mm (17·5 in.). This last appeared to be a brown trout not a sea trout.

Anglers fishing in the tideway occasionally caught trout, and once the 'cleaner' Thames became newsworthy they were increasingly often reported. Some of the earlier and more interesting records are cited here. In January and December 1971 two small trout of estimated weights 113–142 and 227–283 gm (4–5 and 8–10 oz) were caught at Strand-on-the-Green, downstream of Kew Bridge by anglers (C. R. Colman – personal communication). In October 1971 a 566 gm (1 lb 4 oz) trout was caught at Richmond (*Angler's Mail*, 9 October 1971). In May 1972 two trout were caught at Chiswick Bridge which measured 237 and 212 mm (9·3 and 8·4 in.), both appeared to be two winters old when caught, but the smaller fish had many damaged scales on its body which made age determination difficult. Five trout were caught by anglers at Strand-on-the-Green in the summer of 1975, the largest of which were 680 and 340 gm (1 lb 8 oz and 12 oz) in weight (C. R. Colman – personal communication). More recently, a small brown trout 178 mm (7 in.) body length was captured on 22 May 1978 on the foreshore at Greenwich by an angler using maggots as bait. Of the fish which I examined, those caught at Chiswick Bridge had been feeding on small roach and bleak, but the smaller had numerous spiny-headed worm parasites in its lower gut which showed that in earlier life it had fed on gammarid crustaceans (freshwater shrimps) which are the first host of the parasite. Both the 1971 and 1972 fish caught in the Wandsworth area had also been feeding on young bleak. It might be suggested that the abundance of small cyprinid fish like the bleak and roach is probably responsible for the large size and fast growth rate of some Thames trout. Fish featured largely in the stomach contents of three trout from the non-tidal river on 27 June 1971, 230 mm (9·1 in.) body length from Kingston Bridge; 30 July 1971, 520 mm (20·5 in.) and weight 2·718 kg (6 lb) from Laleham; and 20 February 1972, 365 mm (14·4 in.) and weight 1·246 kg (2·75 lb) from Hampton Court.

There are very few literary records of sea trout in the Thames from the nineteenth century or earlier. Buckland (1879) claimed that he received the bodies of sea trout from the estuary in most years at that time, and Murie (1901) reported that they occurred in the lower Thames on occasions (and were sometimes confused with salmon). His claim (Murie, 1903) that two small trout between Waterloo and Hungerford Bridges in 1880 were sea trout, although accepted by Solomon (1976), seems unlikely in view of their size, for they were only 247 and 343 mm (9·75 and 13·5 in.) in length.

On account of the sparsity of published information on the Thames sea trout it is of interest that a considerable number have been recorded since 1971. The first fish to be recognised as a sea trout was found stranded but alive on a mudbank at Deptford Wharf, Charlton on 9 February 1971. It was 762 mm (30 in.) in length and weighed 1·471 kg (3 lb 4 oz). It was a female fish with large eggs in the ovaries and clearly had not spawned during the past winter. Unfortunately, it has to be admitted that this fish was in very poor condition and weighed approximately a third as much as a fish of this length in good condition should do. Probably it had entered the Thames to spawn but had failed to find a suitable spawning ground and had starved while in the part of the river which was still in poor state. As it happened it also was probably not the first sea trout in the tidal river! Fishing at Teddington Lock a member of the Francis Francis Angling Club captured a sea trout in December 1970 of 1·345 kg (2 lb 15 oz 8 dm) which was identified as usual in that club by the officers, before it was released. In view of the next capture there seems no reason to doubt the authenticity of this fish. In March 1971 a fine, fat sea trout of 373 mm (14·7 in.) body length, and weight of 1·096 kg (2 lb 6 oz) was caught in the weir-pool at Teddington. A female fish, it had been feeding on bleak (at least three fish), and maggots on which it was caught. Scale readings showed that this fish had spent only one year in the sea before returning to the river, the Deptford fish had spent two years in the sea; both had migrated to sea after two years of freshwater life. In contrast to the Deptford fish the Teddington trout was in excellent condition and had obviously fully recovered from spawning in the winter of 1970.

In succeeding years several sea trout were captured and the salient details of those that were personally verified are listed below.

1971	October	Tilbury power station; 508 mm (20 in.)
1972	10 January	near Greenwich; female, 560 mm (22 in.), 1·472 kg (3 lb 4 oz)
	10 July	Northfleet, near Gravesend; female 595 mm (23·4 in.), 3·992 kg (8 lb 13 oz)
1973	12 January	Tilbury power station; male 620 mm (24·4 in.)
	19 September	off pier Southend-on-Sea; 1·642 kg (3 lb 10 oz)
1975	30 January	Tilbury power station; female 430 mm (16·9 in.)
	25 February	Tilbury power station; female 330 mm (13 in.), 806 gm (1 lb 12 oz)

	23 November	below Teddington weir, 1·094 kg (2 lb 6 oz 10 dm)
1977	9 October	Thurrock Yacht Club; male 546 mm (21·5 in.), 2·039 kg (4 lb 8 oz)

There was a report of an 3·624 kg (8 lb) sea trout in the Royal Albert Dock, North Woolwich, found when the dock was drained in November 1977, but newspaper photographs showed a heavily spotted fish which looked more like the large Thames brown trout found commonly upstream.

It seems clear from these isolated captures (and I have been informed that through 1976 and 1977 Tilbury power station caught a number of sea trout) that in Gravesend and Northfleet Reaches sea trout are relatively frequent in occurrence, especially in the autumn and winter. The occasional specimens which are captured further upstream shows that there is a tendency for them to move into London, and suggests that there is already a nucleus of Thames sea trout in the river. It seems likely that within the next decade a sea trout population will become established unaided; this would be a major advance in the recovery of the fauna and a striking example of the success of the cleaner Thames.

The rainbow trout, *Salmo gairdneri*, is a native of the western coastal region of North America, and has been imported to Europe and many other parts of the world for its angling and food value. Several reservoirs and other fisheries in the London area are regularly stocked with rainbow trout, for despite its wide artificial distribution it has established very few self-sustaining breeding colonies in the British Isles. It is slightly more tolerant of high temperatures and low dissolved oxygen than the native brown trout which, while adding to its appeal as a subject of intensive fish farming, also makes its occurrence in the tidal Thames less surprising.

It has been captured in the tideway on a number of occasions between 1968 and 1977, but there was a sudden increase in the number reported in the latter year. The details of captures are set out briefly below; further details were given in Wheeler (1978c).

1968	December	Richmond; weight 680 gm (1 lb 8 oz)
1972	18 January	Fulham; fork length 356 mm (14 in.)
1975	17 April	Brentford; total length 520 mm (20·5 in.), weight 1·580 kg (3 lb 5 oz)
	28 May	West India Dock; fork length 292 mm (11 in.)
	15 October	Millwall Dock; two, fork lengths 330, 343 mm (13, 13·5 in.)
1977	4 June	West India Dock; fork length 220 mm (8·7 in.)

8 June	Leigh Creek, Leigh-on-Sea; fork length 270 mm (10·6 in.)
July	Sunbury, Middlesex; weight 963 gm (2 lb 2 oz)
21 July	20 m from Gravesend ferry, Kent; body length 285 mm (11·2 in.)
2 August	Tilbury Dock, Essex; fork length 400 mm (15·8 in.)
10 August	Putney; body length 325 mm (12·8 in.)
11 September	foreshore at Royal Naval College, Greenwich; fork length 318 mm (12·5 in.).

In addition, between mid-June and early December 1975 no fewer than 12 rainbow trout were caught along the 200 yards of Strand-on-the-Green foreshore below Kew Bridge; they weighed between about 0·453 and 1·246 kg (1 lb and 2 lb 12 oz) (C. Colman – personal communication).

Additional to these is the capture of a 0·637 kg (1 lb 6½ oz) rainbow trout by an angler using worm as bait on 23 May 1978 in the Regent Canal Dock which lies 4 km (2·5 miles) below London Bridge (C. Sweet – personal communication).

It must be concluded that there had been a considerable escape of rainbow trout from a stocked water into the tidal Thames; from whence they came is unknown, although I have received reports that they have also been caught by anglers in the lower River Lee, which suggests that they may have escaped from the stocked fishery at Walthamstow Reservoirs. It is interesting also that several of these fish have been taken relatively far down the tidal river, at Gravesend, Tilbury, and Leigh Creek, for the rainbow trout rarely enters the sea in Britain. All these Thames specimens were in very good condition, the Gravesend one being remarkably fat. Most contained gammarid (freshwater shrimp) remains, but the 1975 West India Dock specimen had its stomach filled with *Daphnia* (water fleas), *Gammarus salinus*, and elvers about 63 mm (2·5 in.) in length. It is obvious that the tidal river offers a very suitable habitat to this introduced trout.

The smelt, *Osmerus eperlanus*, although not strictly a member of the salmon family is nevertheless a close relative. Like the trout and salmon it has a small fleshy (adipose) fin on the back between the dorsal fin and the tail fin, and like the salmon and the sea trout it is migratory, breeding in freshwater or where the final effects of the tidal influence is felt. It is a relatively small fish, growing to an absolute maximum of 450 mm (17·7 in.), but the largest fish rarely exceed 305 mm (12 in.) and these are all females. Its two most striking characteristics are the strong scent of cucumber that the fish possesses and the large number of needle-like teeth it has in its jaws and mouth generally.

In the Thames, as we have already seen, the smelt was a very valuable food fish and its remains have been found in several London medieval archaeological sites, so evidently its exploitation has been for long established. There was no evidence to suggest that during the serious pollution of the river in the mid-nineteenth century and again in the mid-twentieth century the smelt survived in the river, although in the outer estuary, as in the Medway, it was common until the 1930s, and in the Crouch and Blackwater mouths the species was relatively common in the 1950s and 1960s, and must have bred in these short rivers. In the tidal Thames the smelt was captured almost from the earliest days of the 1967–73 survey, and it was fascinating to see the numbers increase every year.

Upstream, at the Wandsworth Bridge area, smelt were first captured between April and July 1968, one fish 163 mm (6·4 in.) body length, but the first capture in the river as it began to recover was made in April 1966 when a fish of 143 mm (5·6 in.) was caught at West Thurrock, near the Dartford Tunnel. During the power station survey of 1967–73 smelt were captured at all the stations on the Thames. The number of fish caught in each year and at each area are given in Table 6.1. This shows a fluctuating but generally steady increase through the six-year period, from nil fish in 1967. It also demonstrates that while smelt occurred with some frequency towards the mouth of the river from 1968 onwards, they were penetrating upstream to the west of London in that year and continued to do so each year thereafter. Their apparent absence in the Wandsworth Bridge area in 1973 is not significant as, of the two power stations used at this area, one had closed down and the other was out of commission during the critical spring period.

TABLE 6.1 *The numbers of smelt captured at power stations in four areas of the tidal Thames 1968–73*

	Wandsworth Bridge	*Blackwall Tunnel*	*Outfalls*	*Dartford Tunnel*	*Total*
1968	1	1	—	4	5
1969	1	3	—	2	6
1970	2	9	—	4	15
1971	1	4	—	6	11
1972	3	3	1	5	12
1973	—	—	1	4	5

Later occurrences of the smelt have shown that the species has become extremely common in the estuary. Specimens captured by Mr M. Andrews of the Thames Water Authority have been made available to me and for this I must take the opportunity of acknowledging my thanks to him and his capable team of biologists. Captures of smelt after 1973 increased notably, in 1975, Blackwall Tunnel area (13), Outfalls area (1), Dartford Tunnel area (35); in 1976, Blackwall Tunnel area (23), Dartford Tunnel area (124); and from January to March 1977, Dartford Tunnel area (61). In October and November 1977 during a six-hour sampling period, totals of 324 and 319 smelt weie caught at West Thurrock generating station. These figures show most dramatically how the smelt population in the tidal Thames has increased. They also demonstrate the accuracy of the prediction I made in 1969 that 'it seems clear that there is a sufficient population of smelt in the Thames estuary . . . to give rise to a smelt run up the Thames'; a prediction challenged by Huddart and Arthur (1971), whose contribution to the study of fishes had hitherto escaped the attention of ichthyologists and whose knowledge of Thames fishes was confined to sampling at one powei station during a three-year period.

Andrews (1977) has demonstrated how abundant the smelt was in the upstream tidal river during the extreme drought of 1976, and drew attention to the presence of 'recently hatched fishes taken in June during trawls at Greenwich and London Bridge'. He captured smelt by trawling at Greenwich, Limehouse Reach, and Tower Bridge, as well as in Chelsea Reach. During 1976 smelt were captured by anglers at several points along the river; at a dock at Tooley Street, Southwark (October), at the mouth of the Beverley Brook, at Barnes (October), and in Shadwell basin regularly from August onwards. The abundance of smelt upstream was tragically demonstrated in late August 1977 when, following stormwater discharge from rain-filled sewers, between 30 and 40 dead smelt were recovered from the river at Woolwich (A. E. Hodges – personal communication).

The large numbers of smelt found in the Dartford Tunnel area in the autumn of 1977 were mostly between 75 and 100 mm body length (3–4 in.) and examination of their scales showed that they were in their second year of life (i.e. they had been hatched in 1976). As the smelt spawns in March–April (Belyanina, 1969) they must have been just over eighteen months of age. The 1976 year class seems to have been exceptionally large, and possibly marked the fully successful establishment of a smelt run in the river. In late June 1976 two fish of the year were caught in the Blackwall Tunnel area, 47–52 mm (1·9–2·1 in.), and later catches in Chelsea Reach in December, and at Charlton in November also revealed numbers of young fish of 78–94 mm and 70–93 mm body length (3–3·7 and 2·8–3·7 in.).

The biology of the smelt in the Thames is far from clear. Amongst the November 1977 fish those below 85 mm (3·4 in.) were sexually immature, but both males and females longer than this contained ripening sperm and eggs (and would thus have spawned the next year). Similar observations have been made of November- and December-caught fish in the Blackwall Tunnel and Charlton areas. However, large females which were fully ripe have been caught in March and April at West Thurrock (215 and 196 mm (8·5 and 7·7 in.)), and off Southend-on-Sea in mid-February 1977 230 mm (9·1 in.), in the former case (29 April 1974) with both smaller males and females which had already spawned. This observation is at variance with those of other workers elsewhere in the northern hemisphere who have observed that the size of smelt decreases during the spawning run, the larger fish spawning first (Belyanina, 1969). It may be, of course, that these large, late-maturing females do not spawn in the upper tideway, but belong to tributary spawning stock.

So far as can be established there are large numbers of immature smelt in the whole of the lower tideway throughout the year. The spawning migration upstream reaches the Blackwall Tunnel area in December and potential spawners can be found there through to February; mature and immature fish have been found upstream in the centre of the city in December. The presence of spent fish (i.e. those that have spawned) at West Thurrock in late April suggests that spawning takes place in March, which is borne out by the capture of young fish of the year in June. Spawning must be followed by an immediate return to the lower estuary.

It is not known where the smelt spawn in the Thames but to satisfy the usual requirements of the species it must be upstream of Wandsworth Bridge. Normally, their eggs are shed on stones, water plants, submerged bushes, grass and other solid structures (Belyanina, 1969); they never spawn on mud. Such requirements therefore seem to preclude spawning in the metropolitan and lower reaches of the river.

The return of smelt to the tidal Thames in such large numbers is the most impressive proof of the improvement in the condition of the water. There is now clear evidence of spawning migration upstream and the successful run to the sea of the young fish. No other species so clearly demonstrates the success of the various authorities in controlling pollution. Moreover, it is a result which has direct benefits to man, for the smelt, although small, is now often caught by Thames anglers, and soon there will be an ample stock in the river for a commercial fishery for these delicious and highly priced food-fish. Londoners may soon have the experience, denied them for more than a century, of eating fresh Thames smelt!

STICKLEBACK FAMILY

The three-spined stickleback, *Gasterosteus aculeatus*, is an abundant fish through the coastal regions of Europe, northern Asia, and much of North America. In the southern parts of its range it is largely a freshwater fish, found in lakes, streams and rivers, more or less wherever one looks for it. To the north, as in Scottish waters, it is also found in the sea and breeds in tidal pools, and is occasionally captured far out to sea. Changes in salinity therefore are of little importance to the species, although the temperature of the water makes a difference to its sea-going tendency. However, it is sensitive to low levels of dissolved oxygen (much more so than its relative, the nine-spined stickleback), and although it has a broad tolerance of changes in dissolved oxygen levels it will not live in heavily polluted water.

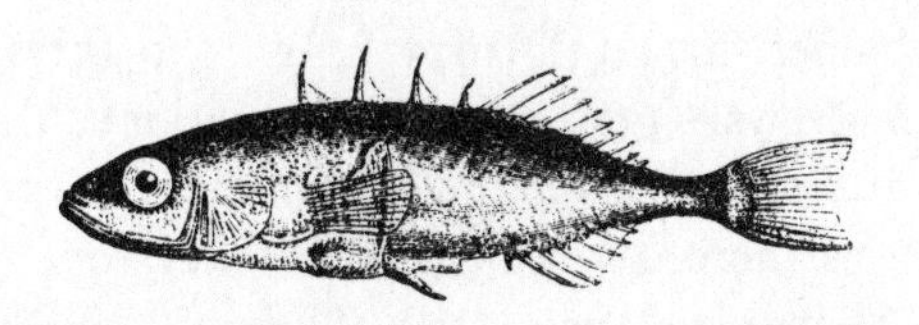

FIGURE 17 *Three-spined stickleback*

In the tidal Thames the stickleback was abundant upstream in the Richmond area (Wheeler, 1958) even when pollution of the metropolitan reaches was severe. During the 1967–73 power-station survey its occurrence, and increasing abundance in central London, was demonstrated from the first years of collection.

The total captures set out Table 6.2 give a general picture of the abundance of the stickleback in the river, but must be interpreted in the light of the methods used to save the fish. The apparent sudden increase in the Outfalls area in 1972 was due to the participation of Ford's Works, Dagenham, in the survey, for this station appeared to be uniquely

TABLE 6.2 *The numbers of three-spined stickleback captured at power stations in four areas of the tidal Thames*

	Wandsworth Bridge	*Blackwall Tunnel*	*Outfalls*	*Dartford Tunnel*
1967	14	7	3	—
1968	14	30	3	1
1969	45	42	1	6
1970	22	13	—	1
1971	86	18	—	—
1972	31	15	60	1
1973	3	2	6	—

equipped to catch small fish by reason of its position; it does not therefore indicate a striking increase in the abundance of the species. At Dartford Tunnel also, the relatively few sticklebacks caught were certainly an under-representation of the population in the river due to the small size of this species and the often large bulk of fish and other animals caught. Huddart and Arthur (1971) reported their capture there in November 1968 (2), February and December 1969 (1 each), and February 1970 (1); these figures supplement my own captures for this region. Later collections using slightly different methods produced much greater numbers here, November–December 1974 (3), 1975 (28), 1976 (51); January–March 1977 (30) (M. Andrews – personal communication). Mr Andrew's work at the upstream power stations in the Outfalls area showed how abundant the stickleback was there; November–December 1974 (17), 1975 (448), and January to September 1976 (115). The great majority of these fish were again caught at Ford's Works, Dagenham, and included catches of 148 fish and 87 fish in July and November 1975.

Netting in the upper tidal river showed that the stickleback was equally abundant there. Hauls of the net in Corney Reach, Mortlake, and Syon Reach in May 1971 yielded an average of 17·3 fish per sweep, and in July 1972 at Syon Reach and Richmond an average capture of 38·2 fish was obtained. These captures and those made at power stations between 1967 and 1973 show that this fish was a common inhabitant of the Thames during this period, being most abundant upstream and least abundant at the mouth of the river, and probably increasing in numbers as pollution of the river lessened.

Study of the lengths of the sticklebacks in the river showed two interesting trends (Table 6.3). First, the number of large fish was greater in the Blackwall Tunnel area than further upstream. Second, the fish captured at Dagenham were notably smaller on average and in size range. This was probably the result of the siting of Ford's Works, Dagenham, close to the confluence of the River Beam and the Thames, which means that the fish sampled originated in the smaller tributary, and were not strictly Thames fish at all. It must be assumed that the stickleback was breeding in the Beam and young fish were moving downstream towards the Thames and capture. The apparent increase in size at the Blackwall Tunnel area may be due to enhanced growth in this region (compared with upstream), although in view of its proximity to the zone of least dissolved oxygen, this seems unlikely. It might possibly be due to penetration of sea water in this region (sticklebacks found in coastal waters of the central North Sea tend to be larger than freshwater specimens' maxima), but again this seems unlikely. A more probable explanation is that the Blackwall Tunnel area fish were recruited by downstream drift from the upper reaches of the Thames and consequently the population consisted of older, larger fish than that upstream. It follows from this that the Blackwall Tunnel area fish at this period were probably not breeding, the muddy bed and absence of vegetation and the tidal nature of the river being unsuitable for the nest-building of this species. This would account for the sparsity of young fish in the area. However, several heavily gravid females with bodies distended with eggs were found in the April 1972 sample at the Blackwall Tunnel area, although this does not necessarily mean that these fish would have bred, only that they could have done if conditions were suitable. The relatively few sticklebacks listed in Table 6.3 from the Dartford Tunnel area were again all large and support the suggestion that they had not bred in the area. It is possible that most of the sticklebacks found in the lower tidal Thames are expatriate losses to the breeding stock.

TABLE 6.3 *Occurrence of sticklebacks in 5 mm length groups (body length) in the tidal Thames*

Length (mm)	16–20	21–5	26–30	31–5	36–40	41–5	46–50	51–5	56–60	61–5	66–70
Syon Reach (May)	—	—	—	—	4	13	13	6	—	—	—
Syon Reach (July)	16	41	32	7	3	20	44	30	13	3	—
Mortlake Reach	—	3	—	1	—	12	8	4	—	2	—
Wandsworth Bridge area (April–July)	—	—	5	46	57	14	5	9	—	—	—
Blackwall Tunnel area (all seasons)	—	—	1	2	4	8	14	13	7	3	1
Outfalls area (all seasons)	—	2	2	4	2	8	—	—	—	—	—
Dartford Tunnel area (all seasons)	—	—	—	—	1	4	2	2	—	—	—

Despite this, the three-spined stickleback is extremely abundant in tributaries of the main river and in still waters adjacent to it. One most interesting temporary population was found in the wholly artificial channel known as the 'King's Scholar's pond', just west of Vauxhall Bridge. This channel is the remaining open course of the old River Tyburn, but now acts only as a storm water channel to relieve the sewers after heavy rain, from whence it is discharged to the river. In July 1969 a total of 81 fish between 15 and 24 mm (0·6–0·9 in.), together with several aquatic invertebrates, were captured in the standing water of the channel. Their origin is a complete mystery; it seems incredible that they were the progeny of adult sticklebacks swept through the sewers by rainwater although that seems to be the most probable explanation. Numbers of sticklebacks were obtained from the West India Docks in June, July and August 1970, 63 fish between 26 and 45 mm (1·0–1·8 in.). The River Lee at Hackney, and in the tidal section, contained sticklebacks in 1965 (Meadows, 1971). Considerable collections were made in 1971 in the River Beam at Dagenham, 28 fish between 18 and 47 mm (0·7–1·9 in.), and in the Mardyke at Purfleet, 43 fish from 20 to 49 mm (0·8–1·9 in.).

BASS, GREY MULLETS AND GOBIES

The bass, *Dicentrarchus labrax*, is a large, silvery, spiny-finned fish which is well known as an inhabitant of the coastal waters of the southern and western British Isles, although it is also abundant on offshore reefs such as the Eddystone. It is, however, a fish which is tolerant of changes in salinity (several of its close relatives in North America being freshwater fishes), and the young especially are common in estuaries.

Most of the specimens which were caught in the tidal Thames were taken in the downstream regions, as one might expect, for it is basically a marine fish. However, it is a species which might be expected to inhabit only water of good quality (although precise information on its requirements of dissolved oxygen and tolerance of toxic substances is sparse), and the very presence of the species in numbers is an indication of the improved quality of life in the lower Thames. Captures of bass were made in the Blackwall Tunnel area in January 1968 (2), and in December 1975 (9) (M. Andrews – personal communication), and in the Outfalls area in March 1972 (1), while Mr Andrews found specimens in October and November 1975 (1 and 2 fish respectively), February and March 1976 (7 and 3), and in August and September 1976 (20 and 170). Further downstream bass were captured in the Dartford Tunnel area in considerable numbers, as follows: 1967 (3), 1968 (13), 1969 (73), 1970 (3), 1971 (9), 1972 (1), 1973 (3). Later sampling by Mr M. Andrews showed that the species was present in abundance through the autumn, winter, and early spring months of 1975–7, on occasions catches of 330 and 262 fish being made in six-hour samples in September and November 1976. Andrews (1977) reported, without giving details, the capture of young bass at London Bridge and in Chelsea Reach during 1976. More recently, on 26 April 1978 a 660 mm (26 in.) total length bass was caught on the coarse grids at Ford's power station (W. S. Moore – personal communication).

The abundance of bass in the tideway in 1976, and their extreme penetration up it, was undoubtedly due to the drought of the summer of that year which increased the salinity of the water upstream. Climatically more normal years would result in the species occurring less far up the estuary. Moreover, the occasional very large catches of young bass at power station intakes must be due to the habit of this species in forming large schools which are occasionally swept up in the intake current of the power station. This alone shows the danger of using power station captures as a guide to overall abundance in an estuary; the fish are not uniformly spaced in the water mass.

The wintertime migration of young bass into the estuary of the Severn has been observed by Hardisty and Huggins (1975), and their observations, made only for twelve months, show that numbers caught begin to rise in October, reach a peak in November, and then stay at a high level until January. As in the Thames the Severn contains few young bass during the summer months, most of those reported here were caught between September and March, with especially great abundance in September and October. Most of the fish captured were young, for example, samples at West Thurrock in January 1968, 12 fish between 56 and 80 mm (2·2–3.2 in.), and November–December 1969, 30 fish between 51 and 84 mm (2·0–3·3 in.). It seems that the tidal Thames is now an important nursery ground for young bass.

Larger fish are from time to time captured. At West Thurrock captures have included: January to April 1968 a fish of 155 mm (6 in.), June 1975 one 1·812 kg (4 lb), 584 mm (23 in.) long; and at Tilbury power station on 29 October 1972 one 460 mm (18 in.) long bass was caught. Anglers' records, as might be expected, list even bigger fish. In chronological order these are: 1968, September, 2·350 kg (5 lb 3 oz) at Tilbury (*Angler's Mail*, 13 September 1968); 1972, August, 1·444 kg (3 lb 3 oz) at Erith (*Angler's Mail*, 31 August 1972), September, 1·869 kg (4 lb 2 oz) at

Gravesend (*Angler's Mail*, 27 September 1972), autumn seven fish between 1·812 and 4·304 kg (4–9 lb 8 oz) all at Gravesend (*Angling Times*, 7 December 1972); 1973, October, 1·812 kg (4 lb) at Gravesend (*Angler's Mail*, 31 October 1973); 1975, November, 3·851 kg (8 lb 8 oz) at Greenhithe (*Angler's Mail*, 19 November 1975), July, 5·153 kg (11 lb 6 oz) at Tilbury (*Angling Times*, 9 July 1975).

Two species of grey mullet have been found in the tidal Thames, the thin-lipped grey mullet, *Liza ramada*, and the thick-lipped grey mullet, *Chelon labrosus*. The latter species is widely distributed in European seas, but becomes less common in Scottish waters and rarer still to the north, while the former's range extends only as far north as Scotland. Both species live in coastal waters and are particularly common in estuaries, but it is the thin-lipped grey mullet which swims up rivers into freshwater. Both feed on bottom mud and will browse on the fine green algae that grows on pier pilings and hard surfaces.

In the Thames, the thin-lipped grey mullet has, as might be expected, proved to be the more common. It was first reported in the Dartford Tunnel area in December 1969, when 2 fish of 100 and 104 mm (3·9 and 4·1 in.) were caught; 2 more, 103 mm (4·1 in.) were caught in January and a third of 86 mm (3·4 in.) in February 1970. Huddart and Arthur (1971) list the presence of this species in this area without giving details, but their survey continued from January 1968 to March 1970. A larger fish, 197 mm (7·8 in.) was caught in this region in May 1973 and another of 340 mm (13·4 in.) in the autumn of 1974. Since the Thames Water Authority has been sampling fish in this region their records show that the species has occurred with increasing frequency, January–March 1975 (5); September–November 1976 (3); January to March 1977 (23) (M. Andrews, personal communication).

Further upstream the thin-lipped grey mullet occurred at the Outfalls area in June 1972, 26 mm (1 in.) long and in the Blackwall Tunnel area at the end of 1971, a larger fish 315 mm (12·4 in.) in length.

The capture of small mullet of this species mainly in the cooler months of the year is similar to that of Hardisty and Huggins (1975) for the Severn estuary. Most of their fish (which were much more numerous) were caught between November and mid-February and measured from 50 to 110 mm (2·0–4·3 in.) in total length, although they added 'occasionally we have found larger fish up to 300 mm in length', an almost exact parallel with the situation in the Thames.

The thick-lipped grey mullet appears to be much less common in the metropolitan reaches of the Thames or indeed upstream at all, although it is common at the mouth of the estuary. Schools of this species can be seen from Southend pier swimming over the intertidal mud-flats, and it is frequently caught by anglers. A single specimen of 274 mm (10·7 in.) was caught in late October 1973 in the Dartford Tunnel area; since then Thames Water Authority biologists have captured the species in September–November 1975 (3), February (2), September (1) and December 1976 (6) in the same area (M. Andrews – personal communication). Curiously, Andrews (1977) reported that this species was 'frequent' in the Thames in 1976 as far upstream as the Blackwall Tunnel, while the thin-lipped grey mullet was said to be 'present' (fewer than ten captures a year) only as far as the Dartford Tunnel. This is the opposite of the normal positions of the species in an estuary, and seems most remarkable.

The common goby, *Pomatoschistus microps*, is an extremely abundant shallow-water small goby, which is found in intertidal sandy bottomed pools, on muddy saltings, and especially in estuaries. Along the Essex coast it is extremely abundant in such habitats, and reports of gobies on the North Kent marshes in the late 1950s probably refer to this species (Wheeler, 1960). As might be expected, it has occurred in the tidal Thames on a number of occasions and is very common in suitable habitats. An early record was of 10 specimens of 17–23 mm (0·7–0·9 in.) body length caught in Tilbury Docks in early October 1966 by a Mr P. Byron. The captor, a keen bird-watcher, had noticed that terns were diving and catching fish in the dock and with the aid of a small net captured these fish which seemed to be common in the main dock and its branches. In the main river the common goby was recorded only a few times and most of those captures were by the Ford's Dagenham power station, where captures of all fish were strongly influenced by the River Beam. Captures here were all made during the months of October and January, *viz.* October–November 1972 (29), November–December 1972 (2), December 1972–January 1973 (3), October 1973 (63), October to December 1973 (14). The lengths and the number of fish are shown in Table 6.4.

TABLE 6.4 *Frequency of occurrence of length groups in the common goby at Dagenham*

Body length (*mm*)	16–20	21–5	26–30	31–5	36–40	41–5	46–50
No. of fish	6	9	30	42	17	3	4

The abundance of small fish (below 35 mm) in these autumn–winter samples shows that the species had bred locally (probably in the River Beam). The common goby, like most gobies, hollows out a 'nest' under a cockle or other shell, under loose pebbles, or other hard structures; the eggs are laid inside the roof and are guarded by the male. It is an extraordinarily successful little fish which breeds prolifically throughout the summer and is present in huge numbers in the sheltered shallow waters of the Essex coast, as in the Crouch and Roach mouths, the Blackwater and the Colne, and in Hamford Water, in the autumn. With the first frosts of winter their numbers decline dramatically and most of the population migrates into deeper water. This general life-cycle agrees with the observations at Dagenham. The fish between 16 and 40 mm were probably all fish which had hatched out during the summer before their capture; those of 41–50 mm (which were all caught in October) were probably the older, more than one-year-old fish which had already bred.

Occasional specimens of common goby were caught in the Dartford Tunnel area, all in the autumn and winter of 1973 (personal observation), and 1974, 1975 and 1976 (M. Andrews – personal communication). Other specimens were caught in the lower Mardyke at Purfleet in late October 1971 (5 fish 19–30 mm (0·8–1·2 in.)), together with eels, three- and ten-spined sticklebacks. More recently, Andrews (1977) has reported the capture of this little fish as far upstream as Chiswick during the drought of 1976.

It is now very probable that the common goby exists in many shallow water and low salinity habitats in the tributaries of the tidal river and wherever the main river offers suitable conditions. Due to its small size it is very likely to avoid capture and remain unnoticed.

The sand goby, *Pomatoschistus minutus*, is a close relative of the preceding species but is somewhat larger, growing to 95 mm (3·75 in.). It tends to live in rather deeper water, avoiding the intertidal shallows and low salinity favoured by its smaller relative. It is nevertheless abundant in the tidal Thames. In the river it penetrated upstream as far as Chelsea in the summer of 1976 (Andrews, 1977) but the first evidence of its occurrence upstream was obtained in January 1969 when, while at Fulham power station, I picked up a live specimen off the screens. Further downstream, in the Blackwall Tunnel area, it occurred in small numbers in 1968 and 1969, while in the Outfalls area it seemed less common until large catches were made by Ford's Works at Dagenham in 1973. From then on, it has remained a common fish in the catches in autumn and winter in this region. Towards the mouth of the river, in the Dartford Tunnel area, it has occurred regularly in autumn and winter from November 1967 onwards, but in ever-increasing numbers and over a wider period of the year. Nearly 3,000 fish were captured in six hours at West Thurrock in December 1976 (M. Andrews – personal communication). In this region it is probably the most abundant fish, and its capture is no longer confined to the winter months. There seems to be little doubt that it is now living in the Thames in the vicinity of the Dartford Tunnel all year round, whereas in the early years of my study of Thames fishes it was entirely a 'winter visitor', able to tolerate the quality of the water only when it was at its most favourable.

Trawling in the mouth of the estuary in August 1970 and May 1971 from the Barrow Deep to Mucking revealed that the sand goby was abundant at all stations fished and indeed, was often the dominant fish both in terms of numbers and mass. In these collections made on the edges of the Barrow Deep and the Yantlet Sands it was found with smaller numbers of a closely related species, *P. lozanoi*, which is so similar that few workers have been able to identify it successfully. This latter species may penetrate further upstream but has not been recognised there.

THE FLOUNDER

The only European flatfish to be found in freshwater is the flounder, *Platichthys flesus*. Superficially it looks like a plaice but lacks the bright orange spots, is coloured dull olive brown on the back and dead white ventrally, and has a line of small prickles along the bases of the fins. Like other flatfishes found in Europe, it breeds in the sea – in the southern North Sea in mid-January (Russell, 1976). Breeding takes place offshore following the adults' migration out of estuaries. The juvenile fish are extremely common close inshore and enter estuaries at an early age. Their progress upriver, against the flow of freshwater, is effected by selecting tidal currents which will carry them upstream (when they can be seen at the surface), while during the ebb they lie on the bottom out of the downstream current. Their ability to select the 'right' current while only months of age and less than a postage stamp in size is truly remarkable.

The abundance today of the flounder in the tidal Thames is another dramatic demonstration of the improved quality of the water in the river. There is no evidence for the occurrence of the flounder between the 1920s and 1968, and it must be presumed that pollution was so severe that it was unable to ascend the river beyond, perhaps, Canvey Island.

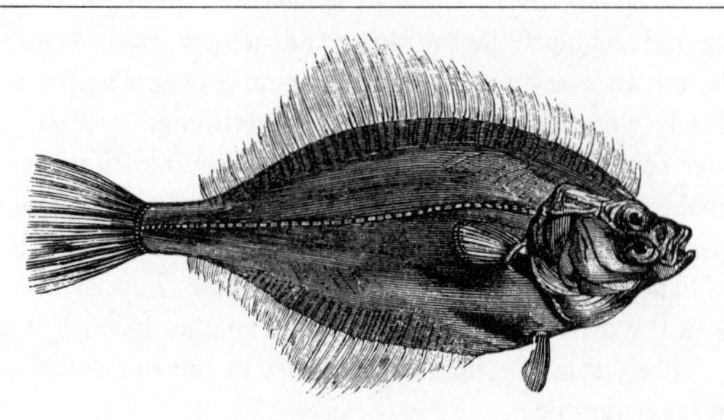

FIGURE 18 *Flounder*

However, in May–August 1969 an 87 mm (3·4 in.) flounder was captured in the Wandsworth Bridge area, the first to have been caught upstream of central London for several decades. Flounders had been caught in the Dartford Tunnel area on numerous occasions from 1967 to 1972 and it was obvious that the species was present there in numbers at least during the winter months. In 1973, however, flounders were present later into the spring and earlier in autumn than in previous years, a trend which has developed into the present resident population of the species in this area.

The captures of flounders at power stations in the 1967–73 survey is shown in Table 6.5. This shows clearly the striking increase in the abundance of the species thioughout the tidal Thames in 1972 and 1973. The 1972 captures at Wandsworth Bridge and Blackwall Tunnel were made chiefly in the period September to December and were mainly of small fish between 60 and 100 mm (2·4–3·9 in.) in length. This suggests that they were young fish of the year (i.e. hatched in 1972) which had migrated upriver during the spring. If this is correct their growth rate was considerably better than the average for the species given by Wheeler (1969).

TABLE 6.5 *Capture of flounders in the tidal Thames 1967–73*

	Wandsworth Bridge	*Blackwall Tunnel*	*Outfalls*	*Dartford Tunnel*
1967	—	—	—	2
1968	—	—	—	1
1969	1	—	—	4
1970	—	1	—	6
1971	—	—	—	4
1972	74	28	12	25
1973	47	20	1	30

The alternative explanation that they were in their second year of life (i.e had hatched in 1971) is not considered tenable because very few flounders were captured in 1971, and, if growth comparable to the average for the species had been made, they should have been much larger than observed. The samples at Wandsworth Bridge in June 1973 containing a majority of fish between 130 and 160 mm (5·1–6·3 in.) support this as these fish are almost certainly the same year class but now in their second year. Because of the interest of the captures of the earliest large number of flounders in the metropolitan reaches of the Thames, the number of fish caught of each length is given in Table 6.6. These figures suggest that the early spring of 1972 saw the first major migration of young flounders upstream beyond the area where dissolved oxygen in the water was low.

TABLE 6.6 *Frequency of occurrence of length groups of flounder in the metropolitan tidal Thames, autumn 1972 to summer 1973*

Length group (mm) >	40	50	60	70	80	90	100	110	120	130	140	150	160	170	180	190
Wandsworth Bridge																
1972																
Aug.–Oct.	1	3	13	17	4	6	3	—	—	—	—	—	—	—	—	—
Oct.–Nov.	—	—	—	—	8	2	8	1	—	—	—	—	—	—	—	—
1973																
June	—	—	—	—	—	—	1	2	2	10	4	12	4	2	1	1
Blackwall Tunnel																
1972																
Aug.–Oct.	—	—	1	5	5	1	6	2	—	—	—	—	—	1	—	—
Nov.–Dec.	—	—	—	—	—	—	2	1	1	—	—	—	—	—	—	—
1973																
Jan.–Feb.	—	—	—	—	—	3	4	2	2	7	1	—	1	1	—	—
March	—	—	—	—	—	—	—	1	2	2	—	3	1	—	—	—

Five young flounders 29–36 mm (1·1–1·4 in.) long were caught while seine-netting at Isleworth on 13 July 1972, which shows both that this year class had moved far upstream, and confirms that they were the fish of the year. A young angler caught a 51 mm (2 in.) flounder at Strand-on-the-Green in late October (C. Colman – personal communication) and in doing so made angling history by catching the first Thames flounder above London for several decades – although his achievement passed unnoticed. By 1973 there were many flounders in the upper tidal Thames and anglers began to catch them in numbers. Unfortunately in June 1973 a number were killed by storm-water pollution in the river, 6 between 128 and 167 mm (5–6·6 in.) from Battersea Bridge, and 5 between 165 and 180 mm (6·5–7·1 in.) from Wandsworth Reach were examined, but eye-witness reports suggested that hundreds had been killed. Anglers' reports of flounders from Eel Pie Island on 19 June 1973, *c.* 152 mm (6 in.) long, and in 14 October from Teddington Weir 97 gm (3 oz 7 dr) (J. Wade – personal communication) showed that these young fish had moved to the furthest extent of the tidal river. A large specimen 343 mm (13·5 in.) in length had, however, been caught at Richmond in November 1972 (*Angling Times*, 26 January 1973).

Later years produced varied further records of flounders in the metropolitan reaches. On 19 June 1975 an observer at Woolwich power station reported shoals of small flounders at the water's edge. Then on 10 July a large number of young fish were caught in Walbrook Dock (a small dock in the City) by Mr A. E. Hodges, Secretary of the Thames Angling Preservation Society. Of the 20 fish examined the length range was 14–40 mm (0·5–1·6 in.) and their average length was 24·7 mm (0·97 in.). Other reports of small, postage-stamp-sized flatfish, such as a number at Barnes Pond, which connects to the river by a pipe, in mid-July 1977 (J. Chambers – personal communication), shows that they are actively migrating up the river during the early summer. Larger fish have also been reported, including a *c.* 305 mm (12 in.) specimen at Teddington Lock on 12 October 1977 (R. N. Green – personal communication).

Downstream of London flounders are a major quarry for anglers, and several localities are singled out as especially good. The foreshore at Tilbury is notable on the north bank and large numbers of fish up to 906 gm (2 lb) in weight are caught there (*Angling Times*, 30 October 1975). At Gravesend the foreshore is again a highly considered area for the capture of flounders. In January 1972 the Gravesend Kingfishers Angling Club held a competition along the promenade and 63 flounders (as well as 125 cod) were weighed in. Large numbers were also captured by anglers at a series of fishing experiments organised by the Corporation of London at Denton near Gravesend between 1973 and 1979. Several specimens between 155 and 290 mm (6·1–11·4 in.) long were examined from the April 1974 catch; all (except the smallest, an immature fish) had recently spawned and all had been feeding heavily on the amphipod crustacean, *Gammarus salinus*, and polychaete worms. Further upstream in the vicinity of the Dartford Tunnel catches of up to 700 flounders have been made in six hours of sampling (August and September 1976) (M. Andrews – personal communication). As previously mentioned, the flounder is now evidently resident all year round in this region but appears to become most abundant from October to January, possibly a result of a spawning migration down to the sea in the late autumn, and a return of spawners and juvenile fish in the spring.

SUMMARY

The return of migratory anadromous fishes such as the smelt and sea trout, and catadromous species, such as eel and flounder, is the most striking evidence of the improved quality of the tidal Thames. The presence of numerous species which are tolerant of changes in salinity along the river, in many cases throughout the year, reinforces this statement. This chapter has been devoted to these migratory and euryhaline fishes which are of the greatest significance in the recovery of the fauna of the tidal Thames. Compared with the huge numbers of flounders, eels, and smelt that the river now holds, the capture of isolated salmon are of minor importance, and plans to restock the Thames artificially are irrelevant. Anglers can now fish all along the river with as good a chance of success as anywhere else, and the huge stocks of smelt and eels offer possibilities of commercial fisheries for these valuable species actually within the estuary. Other potential food-fish which are now abundant in the mouth of the river are discussed in the next chapter. It should perhaps be added that the fish are but one facet of the fauna which has attracted attention, but their presence demonstrates the abundance of food animals that have also returned to the Thames (and fish-eating predators, such as terns, which are now a common sight on the lower tideway). It is the restoration of the fauna as a whole which has greatest significance.

SEVEN

MARINE FISHES

The Thames estuary merges so imperceptibly with the southern North Sea that it is not possible to draw a hard and fast line which delimits the boundary. True, there are administrative boundaries laid down by the Port of London Authority, the City of London's Port Health Authority, and others, but these are entirely arbitrary. From the point of view of the fauna there is a good case for regarding the whole of the Essex coast estuaries and sea inlets (the rivers Roach, Crouch, Blackwater, and Colne, and Hamford Water), as well as the north Kent coast, with the River Medway, as part of the greater Thames estuary. In this area the fish fauna is by and large the same, being composed of species which are well adapted for life in relatively shallow water and on sandy or muddy bottoms, while those elements that are sparse or even lacking are the fishes which favour rocky shores and seabeds or deep water.

For the purposes of this chapter the fishes listed are those which have been caught in the River Thames itself or in the Thames mouth, with references to the status of the species in the greater Thames estuary or North Sea. Species which have been recorded only outside the main river are not discussed. Migratory and estuarine species such as the two shads, the smelt, the grey mullets, the sand goby and the flounder have been mentioned in the previous chapter.

THE LAMPREY, SHARKS AND RAYS

The lamprey, *Petromyzon marinus*, is a jawless fish which obtains its nourishment by sucking the blood from bony fishes. Like its relative the lampern, it spawns in freshwater, migrating upriver from the sea to do so. Because of the widespread pollution of the lower reaches of many of the larger rivers in Europe, navigation weirs and other obstructions, it is now a rare fish in the industrialised lowland regions. Its status in the Thames before the nineteenth century is not well known, possibly it was never very abundant. In more recent times Huddart and Arthur (1971) claim to have captured three specimens in March 1968, which were the first to have been reported this century, and a specimen of 808 mm (31·8 in.) total length was found in May–June 1973, all at West Thurrock power station (Dartford Tunnel area). A fine specimen 670 mm (26·4 in.) in length was caught in Blackwall Reach at the entrance of the West India Dock on 3 June 1978 by a member of the Stepney Angling Club. Occasional specimens have been captured by fishermen in the vicinity of Southend-on-Sea, for example, one of 705 mm (27·8 in.) in length on 26 February 1972 at Shoeburyness.

The smooth hound, *Mustelus asterias*, a small shark with blunt rounded teeth used for crushing the shells of crabs on which it feeds, usually grows to a length of 1·2 m (48 in.). It is relatively frequent in the outer estuary off Walton-on-the-Naze and off the Crouch mouth, but is not often caught close inshore. The only Thames specimen was caught at West Thurrock in October 1973 and measured 300 mm (11·8 in.) in total length. This is the length at which the pups are born, and the West Thurrock specimen may well have been only days old.

The only other shark to be caught during my work on Thames fishes was similarly a newly born spurdog, *Squalus acanthias*. This little fish was only 230 mm (9·1 in.) in total length, and was caught in a trawl in May 1971 on the edge of the Long Sand. In my experience this shark is rare in the outer estuary, an opinion confirmed by Newell (1954) who reported it had been taken only east of Whitstable and then not commonly. The two dogfishes, *Scyliorhinus canicula*, and the nurse-hound, *S. stellaris*, were both reported by Newell in the Whitstable area, and have been occasionally caught in the outer estuary, as in May 1971 on the Sunk Sand, and in the Outer Crouch on a number of occasions, but they do not seem to be very common. Either species is likely to wander into the mouth of the river.

Of the numerous rays found in British seas, the only one to have occurred in the Thames was, paradoxically, one of the rarer species. A small specimen of the sting ray, *Dasyatis pastinaca*, was caught at West Thurrock power station in November 1973; its total length (snout to tip of tail) was 383 mm (15·1 in.), but its body length was only 160 mm (6·3 in.). Again, this must have been a very young fish. Large specimens are, however, common on the Maplin Sands and Dengie Flats in summer and are frequently captured by inshore trawlers and less often by anglers on the Essex coast. Newell (1954) reported that it was not uncommon to find sting rays in summer months in the Whitstable area.

The roker or thornback ray, *Raja clavata*, is extremely common all along the coast at Whitstable (Newell, 1954), and is abundant in summer especially on the Essex coast from Harwich to Southend-on-Sea. Newly hatched young (this species lays its eggs in a hard brown egg case – as opposed to the live birth of the sting ray) and adults of both sexes are caught, and occasional specimens penetrate into the Crouch as far up as South Fambridge. It is therefore rather surprising that this species has not been reported as captured in the lower tidal Thames, but it might be expected to occur.

It is interesting to note here that another rare ray, the marbled electric ray, *Torpedo marmorata*, was caught at Kingsnorth power station in the Medway estuary on 16 October 1973 by Mr W. L. F. van den Broek (personal communication). This species is rare in the North Sea and has not been recorded in the Thames estuary before.

STURGEON

The sturgeon, *Acipenser sturio*, is now an endangered species in western Europe and the nearest breeding population is in the River Gironde, France, so its occurrence within the outer estuary is a matter of considerable interest, as is its occurrence elsewhere in the British Isles. A small sturgeon weighing 30·80 kg (68 lb) and 1·71 m (67 in.) in length was caught on 4 September 1978 near the Tongue light vessel by the Burnham-on-Crouch trawler *Challenger* (E. Edwards – personal communication). While Laver (1898) gave a number of records of occurrence in the nineteenth century of sturgeon on the Essex coast, none were from the Thames. However, a specimen of 27·18 kg (60 lb), 2·134 m (84 in.) was caught at Westminster Bridge in 1867, and Murie (1903) lists captures of sturgeon at Charlton in June 1880 and at Erith in July 1883 while adding that 'many Thames fish of 5, 6 and 7 feet have been noted at various times'.

EELS

Other than the common eel, *Anguilla anguilla*, which has already been discussed at length (p. 000), the only eel to be caught in the Thames is the conger eel, *Conger conger*. It was first caught in November 1967, a fish of 543 mm (21·4 in.) total length, and again in May 1968, two fish 1,035 and 1,105 mm (40·8 and 43·5 in.) long. A fourth specimen was captured in October 1975 (M. Andrews – personal communication). All these fish were caught at West Thurrock power station (Dartford Tunnel area). Several other isolated captures have come to my notice. In December 1972 a 991 mm (39 in.) long conger was caught in the Long Reach at Purfleet (just upstream of the Dartford Tunnel). A 19·932 kg (44 lb), 1,676 mm (66 in.) conger eel was captured at Tilbury power station in May 1971 (*South Eastern Power*, May 1971); a 29·445 kg (65 lb) fish was stranded at Erith the same year (*Angler's Mail* – undated); and a 1,676 mm (66 in.) *c*. 22·650 kg (50 lb) fish was found freshly dead close by Ford's wharf at Dagenham on 6 October 1973 (J. Driscoll – personal communication). The presence of these fish, especially the large ones, is surprising, as the conger eel is, in general, not common in the southern NorthSea.

HERRING FAMILY AND RELATIVES

The herring, *Clupea harengus*, is widespread in British seas, and occurs in the waters of the continental shelf throughout northern Europe and the Atlantic seaboard of North America. Its abundance in the North Sea was legendary (Hodgson, 1957), and its decline through over-exploitation is tragic. Herring were reported in the lower tidal Thames from the beginning of the power station survey (1967–73). In the Dartford Tunnel area, 3 were caught in November–December 1967, another in April 1968 and in early December 1969. In the winter of 1969–70 75 herring were caught, in that of 1970–1 14, in that of 1971–2 75, in that of 1972–3 91. Between January and March 1968, Huddart and Arthur (1971) recorded 41 herring, and later September 1968 to February 1969, 250 were caught. As these authors pointed out, in this period and through to 1972 the herring was apparently absent outside the winter months in the lower tidal Thames. By 1973, however, the season of their occurrence extended from September to May, and this has been confirmed by sampling during 1976 and 1977 by the Thames Water Authority biologists (M. Andrews – personal communication). In these last years, however, the number of fish caught has increased considerably. There is no doubt that now the herring is a common fish in the lower tideway from late autumn to late spring.

It penetrates upstream as far as the Blackwall Tunnel, where small numbers of specimens (1–4) were caught in the first quarter of 1976

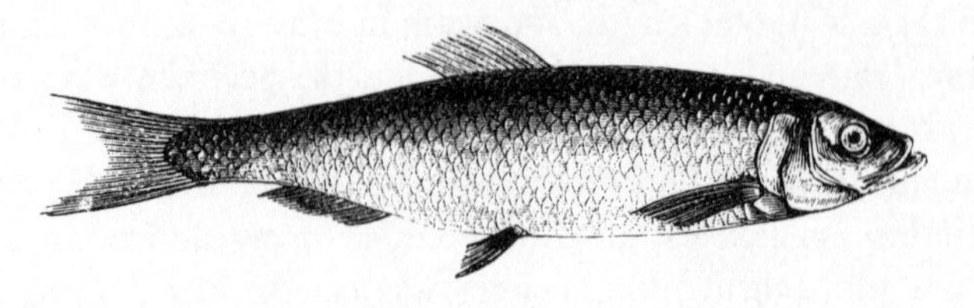

FIGURE 19 *Herring*

(M. Andrews – personal communication). It has been rather more frequently caught in the Outfalls area from December 1972, when the first one was captured. With the exception of this first specimen, which was 203 mm (8 in.) body length, most of the wintertime Thames herring have been small, length range 58–130 mm (2·3–5·1 in.) with the great majority being between 70 and 120 mm (2·8–4·7 in.). Herring of this size are usually found in inshore waters and in estuaries, having been hatched some 10 or 12 months earlier (herring spawn in moderate depths between January and March). Probably most were in their second year but no age determinations were attempted. Some of Huddart and Arthur's (1971) material was examined for them and they reported that the smallest specimens of a January 1969 sample came from offshore stocks while others came from the Channel spawning grounds. They also made the interesting discovery that off West Thurrock some of the herring had been feeding heavily and preferentially on the freshwater cladoceran crustacean (water flea), *Daphnia*.

The sprat, *Sprattus sprattus*, is a close relative of the herring and is widely distributed in the shallow seas of Europe, including the Mediterranean, but which tends to have a more coastal distribution. In the Thames it has always been an important fish, either caught as adults in the mouth of the river or as the dominant constituent of the whitebait further upstream, but, as we have seen, both fisheries were seriously affected by pollution of the river. During the 1967–73 survey it was captured in some numbers in the Dartford Tunnel area, and also, although in smaller numbers, in the Outfalls area and the Blackwall Tunnel area. The total catches are summarised in Table 7.1. These figures are in no sense maxima, for clearly such numerous small fish were difficult to pick up off the screens or out of the trash pit, and many must have been lost. Obviously they were also common and familiar and therefore would tend to be ignored in favour of larger and more unusual species. Despite this there is no doubt that the sprat was common at least in winter in the Dartford Tunnel area as early as 1968, and penetrated in small schools upstream to Blackwall Tunnel. More recent studies by Thames Water Authority biologists have shown that it occurs in small numbers all year round at the Outfalls area, sometimes in larger numbers, for example, 131 fish at Dagenham in September 1975, and also at the Blackwall Tunnel area. Although the numbers fluctuate, and the samples are small, it seems that, unlike the herring, the sprat can be caught in the middle tidal river even in mid-summer. Andrews (1977) showed that during the drought of the summer of 1976 the sprat penetrated upstream nearly to Chiswick and was abundant in the lower tideway.

TABLE 7.1 *Catches of sprat in the tidal Thames 1967–73*

	Blackwall Tunnel	*Outfalls*	*Dartford Tunnel*
1967	—	1	—
1968	—	1	195
1969	16	—	71
1970	1	—	35
1971	12	1	43
1972	—	5	29
1973	—	166	2

Their presence on a year-round basis in the Dartford Tunnel area contrasts markedly with my own observations in 1967–71 when they were caught only between October and March, and those of Huddart and Arthur (1971) who recorded them mostly through the same months. These latter authors reported very large numbers captured on single days (e.g. January 1969 when 2,179 sprats were caught at West Thurrock power station). They also interpreted their catches as showing a migration into the river mouth in August–September of fish of the year (the sprat spawns offshore in early spring and the eggs and postlarvae float shorewards), these young fish (O-group) being captured only in September 1968 (when they were in the 4 to 6·5 cm (total?) length class), and November 1968 (when they measured between 5 and 6·5 cm). This interpretation seems correct, but Huddart and Arthur's other samples seem unlikely to have been correctly aged (unless they were assuming an arbitary birth date of 1 January, which they do not state). In my December 1969–January 1970 samples, the O-group fish are in the body length range of 30–50 mm (1·2–2·0 in.), 1+ fish (i.e. those over a year old are 60–90 mm (2·4–3·5 in.), 2+ fish are 80–110 mm (3·2–4·3 in.). The great majority of sprats caught in the Dartford Tunnel area are between 60 and 90 mm (2·4–3·5 in.) in body length.

The pilchard, *Sardina pilchardus*, is a relative of the herring and sprat, and has a more southerly range, extending from the southern North Sea to the North African coast, as well as the Mediterranean. At one time there were extensive fisheries for the pilchard in the English Channel, and in the southern North Sea (mainly by Dutch fishing boats), but these are of minor importance today; the pilchard is heavily exploited off France and Portugal and sold as sardines. The occurrence of pilchard in the Thames estuary was therefore something of a surprise, as the species is at the northern-most extremity of its range in the North Sea. It has been captured only in the Dartford Tunnel area on 9 and 12 June 1972, specimens 213 and 208 mm (8·4 and 8·2 in.) body length respectively. Newell (1954) reported that odd specimens were taken in trawls at Whitstable, and this serves to confirm the general rarity of the species in the outer Thames estuary.

The anchovy, *Engraulis encrasicolus*, which belongs to another family than the herring, sprat, and pilchard, has much the same distribution as the last species. Before the closure of the Zuider Zee to the sea there was a considerable fishery for anchovies in this vast low-salinity area, but although the fish is still common on the Dutch coast it is not exploited to the same extent (Redeke, 1941). Anchovies were caught on a number of occasions in the tidal Thames; in the Dartford Tunnel area on 9 November 1967 (1) 81 mm (3·2 in.), 2 December 1969 (3) 82, 87, and 97 mm (3·2, 3·4, 3·8 in.), 21 October 1970 (1) 87 mm (3·4 in.), 22 October 1970 (1) 92 mm (3·6 in.), between October 1970 and April 1971 (1) 87 mm (3·4 in.), 10 and 11 June 1972 (2) 131 mm and 96 mm (5·2 and 3·8 in.) respectively, and in October 1972 (1) 75 mm (3·0 in.). Huddart and Arthur (1971) reported the capture of anchovies in the same area in 1969, on 5 November (5), 20 November (4), and 1 December (1). More recent captures at West Thurrock power station have ranged from October 1975 to January 1976, and September to December 1976, with a single specimen caught in June 1976 (M. Andrews – personal communication). Occasionally the anchovy penetrates further upstream; a specimen 75 mm (3·0 in.) long was caught at Barking power station (Outfalls area) on 11 November 1975, and Mr Andrews has reported 3 in the Blackwall

Tunnel region in February 1976. From these records it seems that the anchovy has become considerably more common in the mouth of the Thames in recent years than it was in 1967–9, and the occurrence of specimens in June 1972 and 1976 shows that it is now present at times other than winter months which was the situation in the earlier period.

ANGLER-FISH

The presence of angler-fish, *Lophius piscatorius*, in the Thames was a considerable surprise, for this species is usually associated with open sea coasts. This species has the fascinating habit of living concealed on the sea bed, a nearly perfect match for its surroundings and luring small fish to within snapping distance by wriggling a small flap of skin on the end of a long ray on its back. Its method of attracting prey thus requires clear water for the prey fish to see the lure and the opaque nature of the Thames water would seem far from satisfactory for it. A large specimen, 30·899 kg (68 lb 2 oz) was caught by a human angler from the shore at Canvey Island in 1967; in July 1972 a 1,067 mm (42 in.) fish was stranded at Ford's Works, Dagenham; in June 1975 a 1,150 mm (45 in.) fish was stranded at Cliffe (below Gravesend), and in July 1975 another was stranded on the foreshore at Tilbury. Of these only the first and last were certainly alive at capture but the others must have been living within the Thames estuary. Newell (1954) records the angler as not uncommon in trawls off Whitstable, so presumably these Thames fish must have strayed into the river from this region; it seems to be very rare on the Essex coast.

THE COD FAMILY AND THE VIVIPAROUS BLENNY

As a family the cod fishes dominate the northern European seas, so it is not surprising that more members of the family have occurred in the Thames than any other group of fishes. Several of them are well-known food-fish, such as the cod, whiting, and haddock, but others, such as the rocklings, are small and little known.

The cod, *Gadus morhua*, is widely distributed on the Atlantic coast of Europe and North America. In the North Sea it is more abundant off Scottish coasts than to the south, but in the winter there is a considerable migration southwards, so that large numbers of small to moderate sized cod are caught off southern England, including the Kent and Essex coasts. Not surprisingly, most of the cod caught in the Thames have been taken in winter. The first catch was made in January 1968 in the Dartford Tunnel area when a small fish of 190 mm (7·8 in.) was caught. This was followed by captures in the period October 1970 to January 1971, 1 fish 320 mm (12·6 in.), and in November 1972, 1 fish 590 mm (23·3 in.) in body length. Later captures have continued in increased numbers and frequency, and Mr M. Andrews of the Thames Water Authority has told me of 10 fish in November 1974, 2 in November 1975, 8 in October 1976, 18 in November 1976, 1 each in February and March 1977. The only cod to have been captured further upstream was taken at Barking power station in November 1974. It seems clear that since the early period in the recovery of the Thames cod have increased in numbers considerably.

Large specimens are present in the outer estuary also. Anglers have caught fish weighing up to 9·06 kg (20 lb) at Gravesend (*Angling Times*, 7 December 1972) and again in 1973 (*Angler's Mail*, 28 November 1973), and during that year the total number of cod in excess of 4·53 kg (10 lb) caught there exceeded 450 fish (*Angling Times*, 12 December 1973). Fishing in this area improved so much in 1972 and 1973 that local angling clubs stopped travelling to the Channel coast for their outings and fished from the shore at Gravesend instead. The presence of large fish has continued, a 6·795 kg (15 lb), 864 mm (34 in.) long cod was caught at West Thurrock on 6 February 1977 (E. R. Gill – personal communication). The present abundance of cod in the lower tideway was demonstrated on 22 April 1978 when sixty anglers fished for four hours at Gravesend in the 6th City of London Thames Fishery Research Experiment and captured 133 cod (A. E. Hodges – personal communication). The occurrence of cod in the Thames mouth will no doubt in future be a function of the numbers of cod in the North Sea and climatic conditions, as the fish are no longer affected by the pollution of the river.

The abundance of haddock, *Melanogrammus aeglefinus*, in the northern North Sea and elsewhere was the reason for the invasion of the Thames

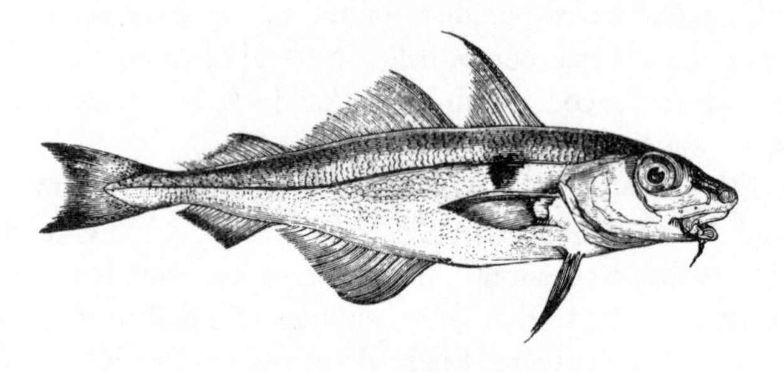

FIGURE 20 *Haddock*

estuary in 1966 and again in 1969. The haddock is not normally found in the southern North Sea today and it is rather uncommon in the English Channel, but in these two years it became abundant. This was attributable to the exceptionally prolific year classes that had been produced three years earlier in the northern North Sea when very many more young fish had survived the critical early months of their life than was usual. As a result of the competition for living-space and food these three-year-old fish had 'flowed-over' from the northern North Sea into the southern part and into the English Channel. As a result haddock were caught in large numbers in places where they were not normally seen.

In 1966 I received reports of over 50 haddock being captured in the Southend-on-Sea area, and others were caught on the Essex and North Kent coasts and on the Channel coast of Kent. All were of this 1963 year class. In the early months of 1969 haddock were again caught in the Thames, but this time far up the river. In January and February 1969 2 were caught at West Thurrock power station (Huddart and Arthur, 1971), while on 3 March 1969 another of 280 mm (11·0 in.) was caught at Greenwich. Examination of the stomach contents of this fish showed that it had been feeding on amphipod crustaceans (*Gammarus zaddachii*), brown shrimps, marine worms and it also contained 3 small sprats. The major invasion of the Thames took place in December 1969–January 1970, when haddock were caught as far upstream as Barking on 6 January, 2 fish 335–56 mm (13·2–14·0 in.). In the Dartford Tunnel area 15 fish were saved for examination between 1 January and 20 March 1970, and many others were taken away for other purposes (they were said to be 'coming up five to ten a day' at Littlebrook power station in January). The fish saved were between 285 and 375 mm (11·2 and 14·8 in.) body length, most were between 350 and 375 mm (13·8–14·8 in.), and were all three-year-old fish.

Anglers caught large numbers in the river, and others were caught when they were stranded in pools on the foreshore at Purfleet. Three of the latter were examined, caught on 10 January 1970, all were maturing females but their stomachs were empty; they measured 360, 380, and 390 mm (14·2, 15·0, and 15·4 in.) body length. Towards the sea more were captured; 40 haddock off Southend in one day (*Angling Times*, 5 February 1970), 3 off Canvey Island, from 0·340–0·906 kg ($\frac{3}{4}$ lb–2 lb) (*Angler's Mail*, 19 March 1970), and a 1·118 kg (2 lb $7\frac{1}{2}$ oz) fish off Benfleet (*Angling Times*, 12 March 1970). The most remarkable capture of all, amongst these relatively small fish, was that of a 3·454 kg (7 lb 10 oz) haddock off Southend (*Angling Times*, 29 January 1970), which was probably one of the survivors of the 1963 year class that had found their way south in 1966. A number of haddock occurred again in the early months of 1971, but they were much fewer in number although larger in size. Two were specially notable, a fish of 2·378 kg (5 lb 4 oz) caught in February, and one of 3·511 kg (7 lb 12 oz) in March.

Unfortunately, following the great haddock years of 1966 and 1969–71 no specimens of the species have been reported in the river. However, their occurrence then showed that pollution is no longer a major factor in keeping them away, and should particularly good year classes arise again in the North Sea then they might be expected to return at least for a short while in winter.

The whiting, *Merlangius merlangus*, is a widely distributed shallow-water species in European seas, the young of which are particularly abundant in onshore waters and in estuaries. The outer Thames estuary has always contained huge numbers of young whiting, especially in the autumn and winter months and thirty-minute tows with a shrimpers' beam trawl frequently yielded up to 50 small fish, all of which were too small to take but which were killed in capture. This was true of the North Kent coast and off the Essex coast, especially within the Crouch and Blackwater estuaries. Its occurrence within the Thames was first noted at the Dartford Tunnel area when one was captured in November 1967, 105 mm (4·1 in.) in body length. Captures thereafter in the winters beginning 1968, 1969 and 1970 were mostly single fish each month. However, Huddart and Arthur (1971) reported 40 whiting caught on 20 November 1969, so it seems that schools of the fish were entering the river when conditions allowed. From the winter beginning 1971, however, the capture of whiting increased considerably and up to 8 fish were caught and preserved each day in October and November 1971, catches which continued at a lower level through to May 1972, and recurred in the winter of 1972–3.

Since then the whiting has increased dramatically in the lower estuary. Mr M. Andrews (personal communication) recorded the capture of 700 fish in January 1975, and his six-hour samples have exceeded 1,000 fish on eight occasions in 1976, the maximum recorded being 3,349 in January 1976. Mr Andrews has also recorded the whiting in small numbers in the winter months at both the Outfalls area and in the Blackwall Tunnel area in 1975 and 1976. Captures in the Dartford Tunnel area are high between September and April, and the abundance of whiting at these periods coincides with its abundance in the Essex estuaries to the north. A similar migration into the mouth of the Severn from the Bristol

Channel has been reported by Hardisty and Huggins (1975).

The specimens caught during the power station survey of 1967–73, although few in number, appeared to represent at least two age classes. The majority of the fish fell into the length range 100–140 mm (4·0–5·5 in.), while the total range for this group was 65–148 mm (2·6–5·8 in.). However, mixed with these length groups in time were specimens of 240, 274, and 281 mm (9·5, 10·8, 11·1 in.). It is probable that these larger fish were 1+ year old, while the smaller ones were O-group fish (i.e. in their first year of life), but it was difficult to establish age for certain as all the fish had been initally preserved in formalin which destroys the otoliths that can be used for age-determination.

A close relative of the whiting, the bib or pouting, *Trisopterus luscus*, is not fished for commercially as a food-fish even though its flesh is no less pleasant to eat. Like the whiting it is extremely abundant in the outer estuary and always has been. Murie (1903) claimed that it was 'pretty much everywhere in the estuary, scarcer above the Nore, occasional in the Sea Reach' and with the exception of Sea Reach this was the situation in the 1950s and 1960s. In the Crouch mouth and the Blackwater it was an exceptional trawl haul that did not catch between 10 and 50 small bib. Most specimens were between 102 and 178 mm (4–7 in.) in body length with only exceptional fish up to 25·4 cm (10 in.); the species attains 41 cm (16·25 in.).

In the tidal Thames the bib was first captured at West Thurrock in November 1967, 2 fish 143 and 224 mm (5·6–8·8 in.) in body length. Thereafter captures in the Dartford Tunnel area were fairly regular during the winter months, although the number of fish caught were few (21 in December 1969 being the greatest number recorded). Interestingly, this species proved to be the most common of the cod family further upstream in the tideway, it being captured in the Outfalls area in January 1968, 2 fish 170 and 171 mm (6·7 in.), February 1970, 1 fish 177 mm (7·0 in.), March 1972, 1 fish 137 mm (5·4 in.), and November 1973, 1 fish 160 mm (6·3 in.). Later collecting by Thames Water Authority staff demonstrated that this species occurred in small numbers regularly here between November and April, and between February and May 1976 it was caught as far upstream as the Blackwall Tunnel area (M. Andrews – personal communication).

Lengths of fish caught in the Dartford Tunnel region suggested that fish of probably three year classes were entering the river in winter. The lengths of the 93 bib caught in December 1969 and January 1970 were plotted in 10 mm length classes, and there were distinct peaks of abundance between 110 and 130 mm (4·3–5·1 in.), between 160 and 190 mm (6·3–7·5 in.), and between 220 and 250 mm (8·7–9·9 in.). This suggests that these fish were in their first, second, and third years of life. Much larger fish have subsequently occurred, such as the specimen of 350 mm (13·8 in.) body length, which was caught in the autumn of 1974.

The captures of bib in the autumn and winter in the lower Thames parallels the observations of Hardisty and Huggins (1975) for the mouth of the River Severn. Here they demonstrated a peak of young fish coming into the estuary in November, although small numbers were caught between September and February. It is clear as far as the Thames is concerned that the water is now of sufficiently good quality to support this fish during its winter-time inshore migration, and the abundance of brown shrimps and other bottom-living crustaceans provide it with ample food. Like the whiting, the bib now finds the Thames-mouth an acceptable nursery ground.

The poor cod, *Trisopterus minutus*, is a close relative of the bib, but even smaller, attaining a length of 23 cm (9·1 in.) but more usually being around 13 cm (5·1 in.). It is widely distributed in European seas and is especially common in the North Sea, although often not distinguished from its relatives. It was first recognised in the tidal Thames in the Dartford Tunnel region in May 1972, 1 fish 121 mm (4·8 in.) long, and again in December 1972, 1 fish 210 mm (8·3 in.). Since then it has become relatively common in this area between the months of November and April and in each winter 1974–5, 1975–6, and 1976–7 it has occurred in numbers of up to 78 fish (M. Andrews – personal communication). Mr Andrews has also discovered that the poor cod penetrates up river as far as the Blackwall Tunnel on occasions, 19 fish being caught there in February 1976.

The pollack, *Pollachius pollachius*, is a common fish in European seas, but one which shows a strong affinity with rocky shores, reefs, and sunken wrecks. It is a rare fish in the southern North Sea and in the outer estuary, and the only record of its recent occurrence to my knowledge was of a specimen caught in 13–16·5 m (7–9 fathoms) on hard bottom in Sheerness Hole in October 1950. It was therefore a considerable surprise to find two specimens in the Dartford Tunnel area, one captured in December 1971, 125 mm (4·9 in.), and the other during November–December 1972, 306 mm (12·1 in.) body length.

The tadpole-fish, *Raniceps raninus*, is widespread in northern European seas but is in general, little known. It seems to be solitary (whereas most cod fishes form schools), and off Scarborough, Yorkshire, where it is common, inhabits rocky areas in 10–40 m (5·5–22 fathoms) in depth. It reaches the southern extremity of its range in the English Channel. In the outer estuary it is rare, and indeed the habitat preference of the species is not well satisfied in this area. Occasionally specimens are taken in the beam trawl around the coast, as off the Stone Banks buoy, Harwich, February 1954 on hard mud and clay, and in the Crouch mouth in December 1962, and May 1978, but the species is not common. It was paradoxical therefore that of the first fish caught in the Thames at West Thurrock power station two were tadpole-fish, caught 11 March and 31 December 1964, and 116 and 128 mm (4·6 and 5·0 in.) body length. Further specimens were caught in the Dartford Tunnel area in November and December 1967, 118 and 117 mm (4·7 and 4·6 in.) respectively, in December 1969, 60 mm (2·4 in.), and April 1973, 148 mm (5·8 in.). Since then the species has occurred in this area in January 1975 (2 fish), April 1976, and December 1976, single fish each time (M. Andrews – personal communication). The tadpole-fish has also been recorded at the Blackwall Tunnel area between January and April 1969 – a fish of 78 mm (3·1 in.) (and again in February 1976 by Thames Water Authority investigators). The repeated occurrence of this fish in the Thames has been rather unusual, but is probably more a reflection of the little knowledge we have of its biology than of a sudden increase in abundance of the species.

There are several small members of the cod family in British waters which are known as rocklings, a name occasionally abbreviated by anglers to 'ling' to the utter confusion of anyone who knows the true ling! Rocklings all have a long barbel on the chin, and two, three, or four barbels on the upper lip or snout. The three-bearded rockling, *Gaidrosparus vulgaris*, is the largest rockling, growing to a length of 530 mm (20·75 in.), and is widely distributed in European seas, usually associated with rocky areas in moderate depths. It has been reported once from the tidal Thames in the Dartford Tunnel area in March 1976 (M. Andrews – personal communication). The four-bearded rockling, *Enchelyopus cimbrius*, is rather smaller and tends to live on soft bottoms rather than rocks; it too is widely distributed. Again, a single specimen was captured in November 1974 in the Dartford Tunnel area (M. Andrews – personal communication). So far as I am aware neither species has been captured before in the Thames estuary.

The five-bearded rockling, *Ciliata mustela*, is by contrast a relatively common fish in the estuary. In the River Crouch mouth numerous specimens were caught by trawling in 1960–3, as they were in the Blackwater. It is also occasionally found on the shore in pools, and under stones; it is especially common in the pools under Southend pier where the piles enter the mud. In the Thames it has been caught on a number of occasions in 1976, in January, and from October to November; on 30 November 5 specimens were taken in the Dartford Tunnel area, and once, in February 1976 at the Blackwall Tunnel area (M. Andrews – personal communication). It was also caught on the Yantlet sands in August 1970 by trawl, but no especial significance can be attached to this as the species was present in this area in 1950. The few occurrences of this species in the Thames, when it is known to be common in the outer estuary, present something of a puzzle. Either the species has not attempted to colonise the cleaner Thames, or so far the methods used to capture it have proved inadequate.

Even more puzzling is the occurrence of the northern rockling, *Ciliata septentrionalis*, another small species with five barbels on the head and supplementary small barbels on the upper lip. It has occurred only twice in the Thames, first in the Outfalls region in January 1975, a fish of 93 mm (3·6 in.), and again in February 1977 in the Dartford Tunnel area. This species was originally described from Norwegian waters but in 1961 it was reported for the first time in British seas, in the Clyde Sea area. Later I discovered it in the Irish Sea, off southern Ireland and in the western English Channel (Wheeler, 1965). It also occurs on the Yorkshire coast, and in the Dutch Waddensea. In more recent years it has been found to be common in the Plymouth area, and in the mouth of the River Severn, where a total of 67 specimens were captured in ten months (Hardisty and Huggins, 1975) and where 353 were caught by four power stations in the Severn estuary and Bristol Channel between November 1972 and May 1976 (Claridge and Gardner, 1977). Whether this species has actually been extending its range southwards during the last two decades or whether it is simply that it is now being correctly recognised is clearly arguable, but its occurrence in the Thames adds further records and tends to support the former hypothesis.

The viviparous blenny, *Zoarces viviparus*, is a slender-bodied fish with low fins and a broad head. Its common name is misleading as it is not a blenny but belongs to a family of fish most abundant in polar seas, but it is viviparous, the female giving birth to fully formed young of about 4 cm (1·5 in.). In the outer estuary it is not common but is caught occasionally in bottom trawls, in the Crouch and Blackwater mouths, and in Hamford Water. The only record of its occurrence in the Thames mouth of which I am aware was of a specimen of 25 cm (9·9 in.) caught by an angler off

Leigh-on-Sea in April 1977. It might well be expected to occur further upstream as well.

GARFISH, SKIPPER, AND SAND-SMELT

The garfish, *Belone belone*, is a long slender fish which grows to 94 cm (37 in.) in length, has a long, pointed snout, with many-toothed jaws, and is a beautiful greeny-blue above and white below. It lives close to the surface of the sea and comes into inshore waters in summer, probably after a northwards migration. It occurs in the outer Thames estuary in late summer and is occasionally captured by anglers. It evidently breeds in the outer estuary because the highly characteristic young fish, about 20 mm (0·8 in.) in length, have been caught in Hamford Water, and Newell (1954) reported them in June off Whitstable. Two larger specimens were captured in a surface trawl in late August 1970 on the edge of the Yantlet Sands; they measured 198 and 217 mm (7·8 and 8·6 in.). Another specimen was caught by an angler off Sheerness in June 1971; it measured 565 mm (22·3 in.) in length.

The skipper, *Scomberesox saurus*, is a fish of very similar build to the garfish, but has a shorter, more slender, beak and a rather deeper body. It too is a surface-living species, but is confined to the open sea and is rarely found on the coast. From time to time it occurs in great numbers, stranding itself on the coasts of the southern North Sea in December and January, and it is thought that schools have entered the North Sea during summer around the Scottish coast and work their way southwards through autumn to become trapped in the shallow and rapidly cooling southern North Sea. The lowered temperature and shallow, tidal water causes the fish to become stranded in schools. This happened on the Suffolk and Essex coasts in 1959, when numbers were found in the Crouch and were caught by fishermen near the East Cant Buoy (Wheeler and Mistakidis, 1960). A more recent record was of a single specimen caught in the Dartford Tunnel area in December 1971; it was 340 mm (13·4 in.) in length.

The sand-smelt, *Atherina presbyter*, is a small fish which grows to a length of 210 mm (8·25 in.), but is usually around 150 mm (6 in.) in length. It is a common inshore and estuarine fish which probably migrates northwards in summer. Its most obvious features are the brilliant silvery stripe down each side, and its habit of forming close-packed, fast-moving schools. In summer it is common in the sheltered waters of Hamford Water, and probably breeds in certain deep saltings pools on

Skipper's Island. It has been caught in small numbers in the mouth of the Crouch off the Dengie Flats, the Buxey Sand, and Foulness Sand in 1952, 1954, 1957, and 1960–3, mostly in summer. Surprisingly for a fish which was so well distributed in the outer estuary, it was not recorded in the Thames until as late as October 1975, when a 61 mm (2·4 in.) total length fish was caught in the Dartford Tunnel area. Later single specimens were caught here in November 1976 and January 1977 and in the Outfalls region in December 1976 (M. Andrews – personal communication). This apparent scarcity and the period of capture contrast with results in the Medway estuary where numbers in excess of a hundred fish have been caught in a day, and the peak of abundance appears to be between April and September (W. L. F. van den Broek – personal communication).

THE DORY AND THE PIPEFISHES

The dory or John dory, *Zeus faber*, is a most striking-looking fish with a deep body, highly compressed from side to side, large eyes and a huge mouth which gives the fish a lugubrious appearance. Its life style is one of gentle movement, often associated with floating wreckage or rowing-boats, and amongst rocks, approaching its prey head on until it is close enough to snap up in its extensible jaws. It is common to the south of the British Isles, but reaches the northward extremity of its range in Scottish waters. It is a solitary fish which is caught from time to time in the Whitstable area (Newell, 1954), and occasionally in the Crouch mouth. Paradoxically, in view of its general rarity, this was one of the first fish

FIGURE 21 *John dory*

to be caught in the tidal Thames, when a small specimen, 223 mm (8·8 in.), was caught in November 1966 at West Thurrock power station. Since then others have been caught at Tilbury power station in November 1972, 280 mm (11 in.) long; off Canvey Island by an angler in July 1977 (B. MacGregor – personal communication); and at West Thurrock power station in the autumn of 1974, 430 mm (16·9 in.) in body length, and in September 1975 (M. Andrews – personal communication).

Three pipefish species have been caught in the tidal Thames, and by far the most common is Nilsson's pipefish, *Syngnathus rostellatus*, a species which is the most abundant pipefish close inshore on sand or mud around the British coast (Wheeler, 1969). It is a small pipefish, rarely growing longer than 150 mm (5·9 in.), and was for long confused by naturalists with the larger, greater pipefish. In the outer Thames estuary it is common, and numerous specimens have been caught in shrimp trawls in the Crouch mouth and the Blackwater, as well as by push-netting in 300–610 mm (12–24 in.) of water on the Dengie Flats. Not surprisingly, it has proved fairly common in the mouth of the Thames. The first specimen was caught in the Dartford Tunnel area in November 1967; thereafter it was captured there in December 1969, October 1970, April and May–June 1972. All were single fish except for the last capture when 5 fish were taken during the six-week period. Further upstream, in the Outfalls area, Nilsson's pipefish has been caught in September–October 1973 (2 fish) and in November–December 1973 (1). All these captures were of fish between 90 and 128 mm (3·5 and 5·0 in.) in length, the average being 101 mm (4 in.). The only significant feature about their lengths was that the Outfalls area specimens were smaller in length, 90–8 mm (3·5–3·8 in.), average 94·6 mm (3·7 in.), and presumably younger fish. Later captures of this species have been made in the Outfalls area in November 1974, November 1975, March and May 1976 (the latter being of 8 fish), and in the Dartford Tunnel area in considerable numbers in most months of 1974–March 1977, the peak catch being 72 fish in October 1976 (M. Andrews – personal communication). The species is even more abundant in the mouth of the Medway where more than 100 specimens were caught in October 1973 and September 1974 (van den Broek – personal communication).

The greater pipefish, *Syngnathus acus*, is rather less common, but is nevertheless frequently caught both in the Medway and in the outer Thames estuary. As its name suggests, it is larger, attaining a length of 47 cm (18·5 in.). In the Thames it was first captured in the Dartford Tunnel area in April 1966, and later captures were made here in March

1968, February–May and October 1970 (2 fish in the latter month). In 1972 it was caught in January–April (3 fish), April–May (3), October–November (1), while in 1973 in April–May 2 fish were caught. Upstream, in the Outfalls area it was captured only in 1973 in April–May and September–October (1 fish in each period). The lengths of all these fishes ranged from 165 to 416 mm (6·5–16·4 in.), but only one was below 298 mm (11·7 in.) in length. Later records of the capture of the greater pipefish show that it has been caught in small numbers in most months between March 1975 and March 1977 at the Dartford Tunnel area, at the Outfalls area, and in the Blackwall Tunnel area (18 fish in May 1976), but in the two latter areas it is only occasional (M. Andrews – personal communication). Three other specimens around 350 mm (13·8 in.) in length were captured in early November 1975 in the Blackwall Tunnel area (R. Dickinson – personal communication).

Another pipefish recorded on a single occasion in the Thames at West Thurrock power station was the broad-nosed pipefish, *Syngnathus typhle*, caught in November 1974. This species is rare in the outer Thames estuary, although I have caught isolated specimens in the Crouch and Blackwater mouths, but if Laver (1898) is to be credited it was 'very common on our *Zostera*-covered shores. It is taken very frequently by the shrimp- and eel-trawlers'. Presumably, when the *Zostera* (eel-grass) beds were destroyed by disease this fish's major habitat was lost, and it became rare.

The seahorse, *Hippocampus ramulosus*, a fish related to the pipefish, is extremely rare, although not unrecorded in the outer estuary. Two specimens were said to have been captured at Brightlingsea around 1866 (Laver, 1898), and a live one was caught at Whitstable in June 1952 (Newell, 1954). Murie (1903) writing of the 1880s, reported 'five or six score' caught in the East Swin and near the Sunk Light (near the mouth of the estuary) and other records from 1876 of Shoebury sands, and 1900 probably in the same area. The occurrence of a small specimen in the Outfalls area on 11 October 1976 was therefore of considerable interest. It was only 32 mm (1·3 in.) long from crown of head to tail tip, and clearly had not been free-swimming for long. In seahorses, as in pipefishes, the male carries the eggs in a pouch and the occurrence of this small fish in the Thames must have been due to the presence of a 'pregnant' male in the estuary.

GURNARDS AND THEIR RELATIVES

Three species of gurnard are found in the Thames estuary of which the tub gurnard, *Trigla lucerna*, is by far the most common. It grows to a large size, up to 75 cm (29·5 in.) and is dull red on the back and upper sides with large peacock blue, green-spotted pectoral fins. Numerous records of its capture exist in the River Crouch and the Blackwater, as well as off Southend. In the Thames it occurred early on in the power

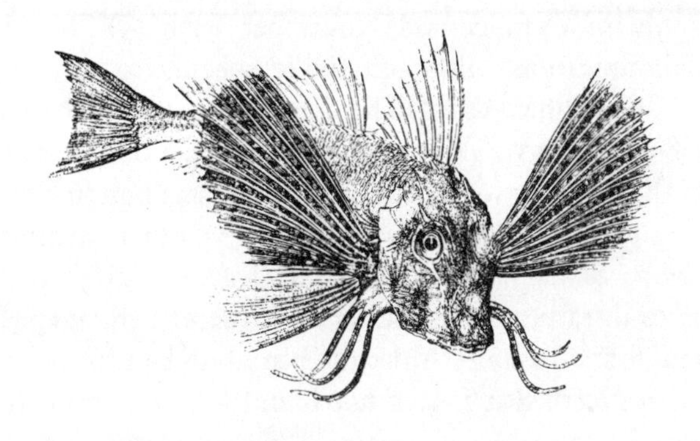

FIGURE 22 *Tub gurnard*

station survey, being first caught in the Dartford Tunnel area in November 1967. Later captures were made here in November and December 1969 (6 fish), October and November 1971 (7), January, May, June, October and November 1972 (6), and January 1973 (2). Further upstream, in the Outfalls area, it was captured in November and December 1972 (2) and between October and December 1973 (2). In the Blackwall Tunnel area one specimen was caught in November 1972. It seems from these captures that the numbers of tub gurnard in the inner estuary were low until 1972 when for some reason it increased and the species both was present in greater numbers through the summer months, and penetrated further upstream, than in previous years. Since the 1967–73 survey this species has occurred in most months of the year through 1975–7 in the Dartford Tunnel area and is caught in some numbers. It has also been captured in the Outfalls and the Blackwall Tunnel area from November 1975 to June 1976 in smaller numbers (M. Andrews – personal communication).

The red gurnard, *Aspitrigla cuculus*, is less common in the outer estuary, although it has occurred on occasions in trawls in the Crouch mouth. In general, it is a rather less common species in British waters than the tub gurnard, is deep red in colour with broad red pectoral fins and is relatively small, growing at most to 400 mm (15·75 in.). Its occurrence in the Thames on a number of occasions is therefore interesting. It has been captured only in the Dartford Tunnel area, and was first caught in August–October 1968. Subsequently, it has been caught in November 1969 (2 fish), October and November 1971 (2), September to November 1972 (3), and September 1973. During the Thames Water Authority work in the area from 1974 to 1977 it seems to have been caught rather less often, only in October and November 1976 (5) (M. Andrews – personal communication), although single specimens were captured in October and November 1977. The lengths of the specimens caught between 1968 and 1973 varied between 103 and 243 mm (4·1 and 9·6 in.) body length, the majority were between 170 and 243 mm (6·7 and 9·6 in.).

The grey gurnard, *Eutrigla gurnardus*, is another small species of gurnard, which, as its name suggests, is basically grey in colour. In general, it lives in rather deeper water than the preceding species and is not commonly found inshore. Although it is abundant in the central area of the southern North Sea I have not found it common in the Thames estuary; Newell (1954) records a single fish at Herne Bay, and single specimens have been caught in the Crouch mouth (1955) and off Sheerness (1950). During sampling in the Thames a single fish was caught at West Thurrock power station in April 1968, 197 mm (7·8 in.) body length. Two further specimens were caught in March and April 1976 (M. Andrews – personal communication).

The bullrout, or short-spined sea-scorpion, *Myoxocephalus scorpius*, is relatively common in the outer estuary and possibly becomes more common in the winter months. I have caught it in some numbers (up to 15 in a haul) in the Crouch mouth in March 1964 and April 1965, but it is more usual to find one or two in a haul. It is also often taken off Sheerness, and off Whitstable. The only occurrence in the mouth of the Thames of which I am aware was of a specimen 140 mm (5·5 in.) caught on 5 January 1972 at Shellhaven, Essex. The reported capture of a long-spined sea-scorpion, *Taurulus bubalis*, of the same length in the Thames below the Dartford Tunnel by an angler (*Angling Times*, 15 June 1977) is noted with reservation as these two sea-scorpions are often confused. This species is, however, known from the outer estuary although uncommon; I have captured it on five occasions in the Crouch mouth between 1957 and 1962.

The hooknose or pogge, *Agonus cataphractus*, is a small fish usually growing only to 15 cm (6 in.) in length. It has a uniquely 'prehistoric' appearance, being covered in hard bony plates while the head has strong hooked spines on the snout and gill covers. It is especially common in the outer Thames estuary in shallow water from Harwich to the North Foreland, and in the Crouch and Blackwater mouths. It is exceptional not to catch several in each haul of the shrimp net in these river mouths. Despite this abundance it was not particularly common in the Dartford Tunnel area, even though it was first caught in late December 1967, in April 1968, and December 1969, a total of 3 fish of 99, 51, and 64 mm (3·9, 2·0, 2·5 in.) body length respectively. Specimens were later caught in November 1974, January 1975, and January to April 1976 in the Dartford Tunnel area, in the Outfalls area in November 1975 and January and February 1976, and in the Blackwall Tunnel area in January, February and April 1976 (M. Andrews – personal communication). This movement well upstream in winter seems now to be well established, as was shown in November and December 1977 when at least 4 hooknoses were captured at Brunswick Wharf power station (Blackwall Tunnel area), and others at Belvedere power station (Outfalls area).

Two species of sea-snail have been captured in the Thames. Of these the most abundant is the sea-snail, *Liparis liparis*, a small fish of 100 mm (4 in.) or so length. It is a soft, gelatinous-looking little fish with small eyes and a powerful sucker disc on its belly with which it normally clings to hard surfaces. In the outer estuary it is caught fairly regularly in small numbers, but on occasions in November-December in the river mouth many may be caught together (possibly as a result of their gathering on suitable grounds to spawn). At the Dartford Tunnel area it was captured in November 1967, again in January 1968 (3 fish) and December 1969 (2). These measured 41 mm (1·6 in.), 61–5 mm (2·4–2·6 in.), and 58–63 mm (2·3–2·5 in.). Since then the species has been captured here in November 1974, January 1975, November 1975 and January 1976, and upstream in the Outfalls area in January 1976, and the Blackwall Tunnel area in February 1976 (M. Andrews – personal communication). Mr Andrews has also told me of the capture of the rarer Montagu's sea-snail, *Liparis montagui*, at West Thurrock power station in March 1976. This fish seems to be very rare in the Thames estuary as a whole.

The lumpsucker, *Cyclopterus lumpus*, is related to the sea-snails and like them has a large sucker disc on its belly. However, its body has a series of hard bony plates on the back and sides, although it is nearly spherical in outline. The species reaches the southward limit of its range in the English Channel and is by no means common in the outer Thames estuary, although Newell (1954) claims that it is 'not uncommon' east of Whitstable. Most of the specimens I have caught in the Crouch and

Blackwater estuaries have been small fish up to 60 mm (2·4 in.). The first specimen to have been caught in the Thames was taken in the Dartford Tunnel area between February and May 1970, measuring 285 mm (11·2 in.). Another was caught in February 1975 (M. Andrews – personal communication).

SCAD, SEABREAMS, RED MULLET, WRASSES, AND MACKEREL

The scad or horse mackerel, *Trachurus trachurus*, is a widely ranging species in British waters and is abundant in schools in the North Sea, but usually offshore. In inshore waters it is not common and in the outer Thames estuary it is very infrequent, although recorded from time to time in both the Crouch and the Blackwater. On the other hand, if a large school of young fish enters the mouth of the estuary (and schools of the magnitude of tens of thousands occur) then the species may suddenly appear to be very abundant. In the Thames the first captures were made

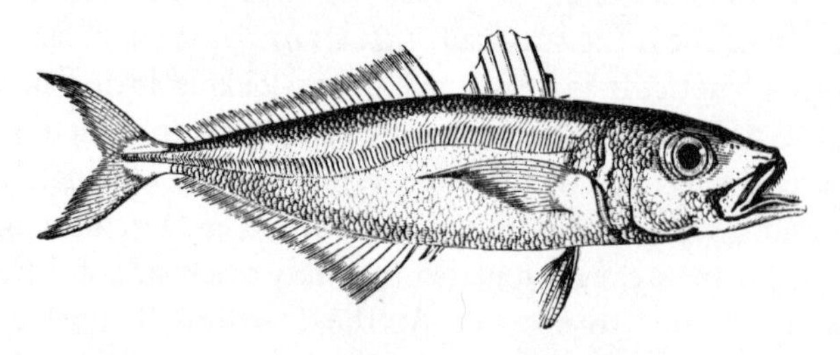

FIGURE 23 *Scad*

on 18 October 1968 when a fish of 290 mm (11·4 in.) length was caught; two 'adults' (size not given) were reported on 22 October 1968 by Huddart and Arthur (1971), probably members of the same school as the earlier one. In November to December 1969 it was again captured in the Dartford Tunnel area (2 fish), then in November–December 1970 (2), November 1971, June 1972 (2), and in October–December 1973 (1). A specimen was also caught in June 1972. Later catches by Thames Water Authority staff show that it still occurred in the area, mostly in September–December, with isolated captures in June of 1975 and 1976 (M. Andrews – personal communication). Huddart and Arthur (1971) also recorded the capture of 114 scad on 5 November 1969 and a further 62 on 20 November. One of the earlier sample was 320 mm (12·6 in.) in total length, the remainder ranged from 71 to 98 mm (2·8–3·9 in.) and the range of the second catch was 77 to 102 mm (3·0–4·0 in.). The two fish

I examined in this month were 85 and 90 mm (3·4 and 3·6 in.) and obviously part of the same school of young fish. In comparing its seasonal incidence with that of whiting Huddart and Arthur were clearly misled by the apparent abundance of scad in November in their samples (which continued for two years only), for the species was not captured again in such large numbers.

At sites further upstream the scad has been captured in the Outfalls area on 19 October 1972 (3 fish) and in October–December 1973 (1). A single fish was caught here in October 1975 by Thames Water Authority staff (M. Andrews – personal communication).

It is a matter of some interest that scad in the Thames have either been small (my samples, like those of Huddart and Arthur already mentioned, ranged from 62 to 90 mm (2·4–3·6 in.)), or much larger (my samples ranged from 200 to 290 mm (7·9–11·4 in.). Clearly two or more age classes are involved in the schools entering the Thames.

The black seabream, *Spondyliosoma cantharus*, is common in the English Channel, a particularly well-known area for it being off Littlehampton where it breeds on the gravel beside hard ledges in moderate depths. It usually grows to a length of around 35 cm (13·75 in.). In the Thames estuary it is rare. Small specimens of 130 mm (5·1 in.) have been captured off the North Foreland and off Margate in November 1960 and Newell (1954) reported a specimen trawled off Herne Bay. Another was caught in the Crouch in January 1967, 290 mm (9·9 in.) in length. In the Thames a specimen was captured in August 1975 at West Thurrock power station (M. Andrews – personal communication); it measured 245 mm (9·5 in.). In 1977 two further specimens were caught in the greater Thames estuary; one on 31 August in a drift-net upstream of Burnham-on-Crouch, 117 mm (4·6 in.); the other in June by an angler off Southend-on-Sea, 225 mm (8·9 in.).

The red mullet, *Mullus surmuletus*, is a deep reddish-brown in colour, and has two long barbels on its chin (they are occasionally called goatfishes on account of these barbels). It is a bottom-living species which travels in small schools and is at the northernmost extremity of its range in British waters. In the southern North Sea, especially on the Dutch coast, it is moderately common. There seem to be few recent records for its occurrence in the Thames estuary, although two specimens were caught off Clacton in July 1950, and its occurrence in the Thames was rather surprising. Nevertheless it has been captured on a number of occasions in the Dartford Tunnel area from October 1970 (3 fish), 155, 165, 175 mm (6·1, 6·5, 6·9 in.), October 1970–January 1971 (1), 178 mm

(7·0 in.), and on 13 June 1972 (1), 112 mm (4·4 in.). The similarity in size of the first four fish suggests that they all belonged to one school. Since then it has been caught in November (1) and December 1975 (3), (and September 1972 (2) M. Andrews – personal communication). A specimen was caught off Canvey Island in September 1977 and was sent to me by Mr B. MacGregor of Southend-on-Sea; it measured 165 mm (6·5 in.).

The ballan wrasse, *Labrus bergylta*, is a species which is mostly associated with rocky areas on the British coast, although it is common in shallow water. It eats shellfish, including crabs, but mainly feeds on mussels. Although Murie (1903) listed two records from the end of the last century, its occurrence in the mouth of the Thames was quite unexpected, but a single specimen, 146 mm (5·7 in.) was captured on 7 January 1975 at Northfleet power station, near Gravesend (M. Andrews – personal communication). Others have been captured by anglers at Southend-on-Sea in July 1977 measuring 205 mm (8·1 in.) and on 12 June 1978, 275 mm (10·8 in.) and weighing 0·708 kg (1 lb 9 oz). Presumably these, like the isolated specimens of corkwing wrasse, *Crenilabrus melops*, captured in the Ray Channel in April 1977, and off Walton-on-the-Naze in June 1978, were the result of postlarval drift from regions like the English Channel where the species is more common.

The mackerel, *Scomber scombrus*, is too well known a fish to require any description but it is usually associated with open sea coasts and moderately deep water. Newell (1954) wrote that it was rare at Whitstable but became more common further to the east, and certainly off the Dutch coast it is moderately abundant at least in summer. It occurs in schools off Southend-on-Sea and is occasionally caught by pier-head anglers. Despite this its occurrence in the Thames is quite surprising. It was first reported to me by Mr J. Driscoll, an experienced angler, who watched a large school feeding on whitebait off Ford's Works, Dagenham, on 16 June 1974. In November 1974 two were caught at Barking power station, and February 1976 one was caught in the Blackwall Tunnel area (M. Andrews – personal communication).

WEEVER, BUTTERFISH, SANDEELS, DRAGONETS, AND GOBIES

The weever, *Echiichthys vipera*, is a short-bodied, rather deep fish, small, at most 14 cm (5·5 in.) in length, but with two strong spines on its head and spines in its dorsal fin which have venom glands attached. Wounds from these spines are intensely painful and produce severe local swelling.

FIGURE 24 *Weever*

The weever's life style is to lie buried in sand with only the top of its head and back visible; it migrates up and down the shore with the tide but is not an active swimmer, nor does it form schools. In the outer Thames estuary it is abundant on the edges of the sandbanks that border the Barrow and Black Deeps, and along the outer Maplin Sands, but close inshore it seems rare. Newell (1954) recorded it on the Tounge Sand (north of Margate) but this again is offshore. It was of some interest therefore when Thames Water Authority biologists recorded its capture on five separate occasions between January and May 1976 at West Thurrock power station (M. Andrews – personal communication).

The butterfish, *Pholis gunnellus*, is a common fish in shallow water and on the sea shore but is usually associated with rocks. In the outer estuary it is not common but is captured from time to time in shrimp trawls as in the outer Crouch mouth in July 1960 and February 1963 and in the Barrow Deep in May 1971. Newell (1954) has recorded it as common in shallow water at Whitstable. The furthest upriver that it has been recorded is at Southend-on-Sea when two specimens were captured in pools at the bases of the pier pilings in August 1973. It may well be reported further upstream in the future.

Three species of sandeel have been recorded in the Thames, but strangely the most common is one which is usually regarded as an off-shore species. All three occur commonly in the southern North Sea, but because of earlier confusion in the taxonomy of the group the status of each species in the outer estuary is not well known. Raitt's sandeel, *Ammodytes marinus*, has been reported on a number of occasions, first at the Dartford Tunnel area in November 1967, and later there in January 1968, and in January (3) and March 1970 (4), February, April and May

1972 (4 total), and March 1973. It was captured in the Outfalls area in November–December 1973 (2). These fish ranged in lengths from 120 to 180 mm (4·7–7·1 in.), but with three exceptions fell into the range of 140 to 166 mm (5·5–6·5 in.). Since then, Thames Water Authority biologists have captured it in the Dartford Tunnel area in January and March 1977, in the Outfalls area in January 1975 (4), January and February 1976 (6 total), and in the Blackwall Tunnel area in January, February, and April 1976 (20 total) (M. Andrews – personal communication). It seems, therefore, that this sandeel is established in the mouth of the Thames and enters the river mainly in winter, in greater numbers and distance upstream under the influence of tide and weather.

The sandeel, *Ammodytes tobianus*, although usually considered to be the more common of the two species, has proved to be less abundant in the Thames. It was first captured in January 1968 in the Blackwall Tunnel area, where it was later taken in February 1969, and in May 1970. In the Outfalls area it was caught in January–April 1968 and in December 1972 (2), while downstream in the Dartford Tunnel area 4 were caught between 19 January and 8 March 1970. These fish measured between 115 and 142 mm (4·5 and 5·6 in.). Later captures of this species have been mainly in the Dartford Tunnel area in January and February 1977 (8 total), but some have been taken in the Outfalls area in January 1975 and in the Dartford Tunnel) area in March 1976 (M. Andrews – personal communication).

The greater sandeel, *Hyperoplus lanceolatus*, was by far the rarest of the three species and was captured only once during my work on the Thames, when in June 1972 a fish of 243 mm (9·6 in.) was caught in the Dartford Tunnel area. Later specimens were caught here in March, June and July 1976 (M. Andrews – personal communication). It is interesting that of these few captures most have been in early summer; this contrasts with the mainly winter-time captures of the other species.

The dragonet, *Callionymus lyra*, is widely distributed in European seas and is the commonest of the three British species. It seems to be most abundant on clean sandy bottoms and is well adapted for lying partially buried in the sand. In the outer Thames estuary it is common on sandy and shell-grounds from Clacton to Sheerness, mostly in depths of 5·5–14·6 m (3–8 fathoms), and there are many records of its capture in the Blackwater and Crouch mouths. In the Thames mouth in May 1971 specimens were trawled on the West Barrow sand in 6–8 m (3·3–4·4 fathoms), on the edge of the Little Sunk Sand and north-east of the West Shingles.

In the Thames it was captured in the Dartford Tunnel on a number of occasions between 1967 and 1973, the first occasion during November to December 1967. Later captures were made in January 1968, December 1969 (10), May 1970, April–May and October–December 1973. A wide range of lengths were involved from 49 to 142 mm (1·9–5·6 in.). Although no age estimation was attempted the length groups fall into three classes, below 59 mm (2·3 in.), between 60 and 99 mm (2·4–3·9 in.), and between 100 and 142 mm (3·9–5·6 in.). This suggests that fish of at least three year groups are present in the lower river and that most are juveniles. Later captures of this species have been made in the same area in small numbers during the winter and early spring in both 1975, 1976, and 1977 (M. Andrews – personal communication).

Three species of gobies have been recorded in the Thames in addition to the common goby and the sand goby which were discussed on pages 161–2. Gobies, as a group, are one of the most successful of all bony fishes and twenty species are recognised in British seas alone. Mostly they are bottom-living fish, some confined to rocky shores, and in general they live in shallow water. Most species have the pelvic fins united to form a disc which has only the weakest powers of adhesion, but can hold a fish in still water against the vertical side of a pool or rock. Curiously, one of the species found in the Thames, and the mouth of the estuary, is a fish normally associated with rocky shores, the rock goby, *Gobius paganellus*. The credit for these captures belongs to Mr M. Andrews and his colleagues of the Thames Water Authority, who first captured this species in Sea Reach in a trawl on 12 September 1975, and later that month found a second specimen in the Dartford Tunnel area while sampling at West Thurrock power station. They found two additional specimens in January 1976 here. These captures closely followed the first capture in the Thames estuary, when a 30 mm (1·2 in.) specimen was captured at Decoy Point, by Osea Island in the Blackwater on 4 April 1974 (Barron, 1976).

The black goby, *Gobius niger*, is not an unexpected inhabitant of the lower Thames, as this species is well known to inhabit water of low salinity. In the outer estuary it is moderately common, but it becomes more so in the Crouch and Blackwater estuaries, although it never attains the abundance of the sand goby. In both these estuaries I have found it in small numbers (1–5) in trawl hauls, but occasionally, presumably where local variation in the bottom of the river makes it an especially suitable habitat, between 10 and 15 will be caught in a twenty-minute tow. In the Thames it was first caught in April 1973, when a single fish,

79 mm (3·1 in.) was caught in the Dartford Tunnel area. Since that occasion it has occurred in May and September 1975 (7 in the latter month), January, March, July to November 1976 (14 total), and January, February, and March 1977 (M. Andrews – personal communication). The presence of the black goby in numbers of this order of magnitude is what one might expect of this species in the river; if anything, it is surprising that it was not captured before 1973.

The transparent goby, *Aphia minuta*, is very small, growing at most to 51 mm (2 in.), it is also one of the exceptional gobies to live in mid-water or at the surface, not on the sea bed. In the estuary in general it is undoubtedly fairly common, but it is extraordinarily difficult to catch without using a special net. Between 1957 and 1963 I recorded it on fifteen occasions in the Crouch mouth using a beam trawl, but these fish were almost certainly captured in the final moments of hauling the net when it was near the surface. In the Thames it was first caught in the Dartford Tunnel area when two specimens, 36–44 mm (1·4–1·7 in.) were caught in May–June 1972. In November 1973 a single specimen was caught in the Outfalls area; this measured 24 mm (0·9 in.). It has been recorded since at the Dartford Tunnel area in December 1975, and in March 1976, when 9 fish were caught (M. Andrews – personal communication).

FLATFISHES

Seven different kinds of flatfish have been reported in the Thames, which means that most of those species found in the outer estuary have at some time or the other been caught. The flounder, which is in many ways the most interesting, has already been discussed. Its near relative, the plaice, *Pleuronectes platessa*, is much less common but nevertheless has occurred in some numbers in the river. In the outer estuary it is very common, but many of the fish caught are small, immature specimens up to 150 mm (5·9 in.) length; despite this very large fish do occur, a 3·659 kg (8 lb 1¼ oz) plaice was caught off Southend-on-Sea pier in March 1976. The plaice has been captured in the Dartford Tunnel area since systematic sampling began in November 1967. In this region it occurred in small numbers in samples taken between October and April each year until 1972. Since then Thames Water Authority staff have taken plaice in most months of the years 1975–7, but there appears to have been a peak of abundance around September and a decline after February (M. Andrews – personal communication). The samples which I collected in this area between

1967 and 1972 contained mostly small fish. The overall length range was from 40 to 160 mm (1·6–6·3 in.), but the great majority were below 100 mm (3·9 in.) body length. It seems from this that, as in the outer estuary, the plaice are young and the whole estuary forms a nursery ground for young fish of this species. This was confirmed by trawling in the mouth of the river in August 1970 and May 1971, when large numbers of young plaice were caught.

The dab, *Limanda limanda*, is a close relative of the plaice, although smaller, and is widely distributed in British seas. If anything, it is more common in shallow water on sandy or shell beaches than the other flatfishes. In the outer estuary it is very abundant in shallow water and it can be caught virtually anywhere between Harwich and the North Foreland in suitable depths. Not surprisingly, in view of the suitability of the general habitat it offers, the lower Thames contains numerous dabs. Again it was first caught in November 1967, and continued to be caught in small numbers through the winter months of 1968–73. On occasions larger numbers were caught, October 1970 (32), November 1971 (18). Again most of these were small fish; the total range in length was 40–150 mm (1·6–5·9 in.), and all but 4 fish fell into two length groups 40–69 mm (1·6–2·7 in.) and 80–119 mm (3·2–4·7 in.). As the species commonly attains a length of 25 cm (10 in.), it is clear that the Thames fishes are mostly juveniles and the river mouth is a nursery ground for this species as well.

The third abundant flatfish in the Thames is the sole, *Solea solea*, a fish which masquerades in fishmongers as 'Dover sole'. It is widespread in North European waters, especially favouring moderately shallow regions where the bottom is sand, fine gravel, or shell, although mud does not deter them. In the outer estuary the sole is very abundant, and appears always to have been so. Murie (1903) referred to it as 'everywhere distributed . . . quite an abundant fish in the Thames mouth from Hole-

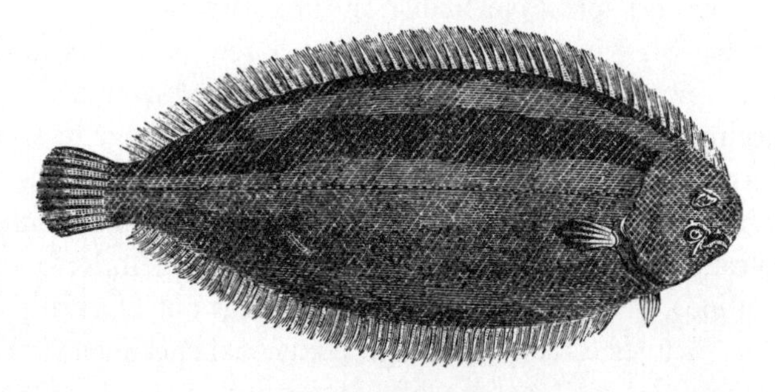

FIGURE 25 *Sole*

haven to the East Girdler', and cited as good fishing areas from Yantlet buoy to the middle Blyth buoy. More recently it has been found to be abundant in the Blackwater and Crouch mouths, and in August 1970 it was caught in some numbers on the West Blyth Sands, Yantlet Sands and even on the edge of the Mucking Flats.

In the Thames itself at the Dartford Tunnel area it was first captured in November 1967, a fish of 103 mm (4·1 in.). Following this it was caught in small numbers in 1968, and 1970 to 1973, although on occasions moderate numbers were collected as in April 1968 (24 in one day) and during April 1972 (34 in the month). Huddart and Arthur (1971) recorded the capture of soles in March, April and November 1968 (23 in four samples) but it did not occur in their 1969 samples until November–December and then only 5 fish were caught. It seems clear that in the 1967–9 period the sole was occurring regularly but not in great numbers in the Dartford Tunnel area. This conclusion contrasts strongly with the data collected by Mr M. Andrews of the Thames Water Authority who, during monthly sampling in 1975 and 1976, has found soles in each sample, sometimes, as in March and October 1976 in large numbers (173 and 152 in six hours respectively). There is no doubt that the sole has penetrated in much greater numbers into the mouth of the Thames between my observations of 1967–73 and those of Mr Andrews of 1975–6.

The soles caught in the 1967–73 sample were mostly small. The total length range was between 50 and 248 mm (2·0–9·8 in.), but only 3 large fish were represented, 205, 225, and 248 mm (8·1, 8·9 and 9·8 in.). The great majority of the samples were between 60 and 109 mm (2·4 and 4·3 in.). As the sole spawns mainly in spring and summer and most of these fish were caught in early spring it follows that they were either at, or nearing, the end of their first year of life. As with the plaice and the dab it is certain that the Thames mouth is an important nursery ground for soles, and from the numbers caught in recent samples this may be the most important species of flatfish in the estuary.

Another flatfish which has occurred on a few occasions is the brill, *Scophthalmus rhombus*. This was first caught in November 1967, and later 3 specimens during October 1971. The length range of these 4 fish was from 102 to 112 mm (4–4·4 in.); thus they were all young fish (the species regularly attains a length of 500 mm (19·75 in.)). Specimens have since been caught in September and October 1976 (M. Andrews – personal communication). From my experience the brill is not especially common in the outer Thames estuary, although occasional specimens are taken in the Crouch mouth and elsewhere, off Clacton and in Hamford Water.

A close relative, the turbot, *Scophthalmus maximus*, seems to be even rarer, but does occur from time to time as was proved by an angler's capture of a 1·246 kg (2 lb 12 oz) fish off Southend-on-Sea (*Angling Times*, 20 May 1971).

Other flatfishes reported in the Thames have included the lemon sole, *Microstomus kitt*, and the scaldfish, *Arnoglossus laterna*. The lemon sole was reported in the Dartford Tunnel area in February 1976 (M. Andrews – personal communication), and in November 1977. Compared with the plaice and dab it is an uncommon flatfish in the outer Thames, but is caught in small numbers, and chiefly young fish, throughout. It seems to be most common in the outer Crouch between August and December, but is present all year round. The scaldfish, which is a very small flatfish, rarely growing longer than 150 mm (5·9 in.), is common in the southern North Sea, but from my experience rare in the outer Thames estuary. The only record of its occurrence was a single specimen, 56 mm (2·2 in.) caught in May 1971 on the west side of the Sunk Sand (at the very extremity of the estuary). It was therefore a considerable surprise when Mr M. Andrews (personal communication) reported the capture of a scaldfish between December 1975 and January 1976 at Blackwall Point power station (Blackwall Tunnel area). How this fish came to be so far up the river remains a mystery.

TRIGGER-FISH FAMILY

The capture of a trigger-fish, *Balistes carolinensis*, in the Thames at West Thurrock power station (Dartford Tunnel area) in the autumn of 1974 was also a considerable surprise because this is a fish which is normally found in the tropical and warm temperate Atlantic Ocean, only making occasional appearances on the western Channel coast and off southern Ireland in late summer and autumn. It was a large specimen, 310 mm (12·2 in.) in body length and was sent to me by Mr E. R. Gill, chemist at the power station. From records received during the course of 1974, however, it became obvious that there had been a large invasion of British seas by this species during the year and numerous captures were reported along the Channel coast and in the Irish Sea, and as far north as Scotland in the west. The Thames specimen must therefore have been part of this invasion, which was probably caused by unusually strong movements of water masses from the subtropical Atlantic, and it was one of the few specimens which passed through the Channel to enter the North Sea. It was an entirely fascinating capture for the Thames.

EIGHT

THE PROMISE FOR THE FUTURE

In previous chapters we have followed the story of the tidal Thames as it has been affected by the human population on its banks, chiefly in the London area. The history of the rich fisheries that existed until the early nineteenth century and their diminution has been sketched, as has the history of the pollution of the Thames. Today, as a result of far-sighted proposals by official committees and firm implementation of measures of pollution control by several authorities the Thames is now cleaner (whatever criteria are employed) than it was a century ago. The return of fish and other forms of wildlife, most notably the waterfowl, as recorded by Jeffrey Harrison and Peter Grant in their fascinating *The Thames Transformed* (1976), has been to some extent an unlooked-for bonus to the major concern for public health risks in the polluted river. Within twenty years the fish fauna has improved from the situation when there were no fish between Richmond and Gravesend, to one in which virtually a hundred species have been recorded, while others are known to be present in numbers in the mouth of the river. This is a situation in which the Port of London Authority, the Greater London Council (and their predecessors the London County Council), and the present Thames Water Authority should take pride, for the restoration of the river is a major success. Paradoxically, the improved quality of life in the river has generated problems which would not even have been considered in the days of severe pollution, for the maintenance and even further improvement of the aquatic ecosystem places constraints on future developments along the river. Some of these will be considered in this chapter.

PENDING IMPROVEMENTS AND POSSIBLE THREATS

The major improvement in the quality of the water in the 1960s and 1970s was affected by a programme of construction of new plant at the major sewage treatment works at Beckton, Crossness, and Mogden, which resulted both in an improved quality of effluent being produced and an increased capacity so as to allow some of the smaller, older, sewage works to be closed down. These works were all within the Greater London Council area. A short account of the other sewage works discharging to the tidal river that lay outside the Greater London Council's area has been given in the Port of London Authority's publication *The Cleaner Thames* (1967). Amongst these was the large works which received sewage from the north-west Kent area, which at that time produced a very poor quality effluent due largely to overloading. This works, then known as West Kent sewage works, was of relatively minor importance to the Thames in its polluted state. However, as the river quality improved, the effect of this discharge became proportionately more serious. The upgrading of this works, now known as Long Reach works, is now underway and was scheduled to be completed by the end of 1978. This is an example of a situation that occurs elsewhere along the Thames; low-quality discharges, which were so small compared with the major sewage effluents that their contribution was of little significance, have assumed a new importance in the context of the cleaner Thames. This may even be true in time of the sewage effluent from Southend-on-Sea which is discharged after only primary treatment through an outfall pipe 2 km (1·25 miles) offshore. In view of the present and probably increasing importance of the Thames mouth as a nursery ground for young food-fish, such as soles, plaice, and whiting, the acceptability of such discharges may be questioned.

The same consideration makes it necessary now to monitor the discharge of toxic effluents, such as the metals cadmium, mercury, zinc and lead, and chlorinated hydrocarbons. A study of the level of mercury in tidal Thames water in 1971 showed that it was low at Teddington and at Southend, but that in the vicinity of the Beckton outfall it was somewhat higher (Smith *et al.*, 1971). However, since that report, the Beckton treatment plant has been considerably extended and there is no reason to suppose that mercury is now at as high a level as it was at the beginning of the decade. In the Medway estuary a more recent study (1973–6) of five heavy metals (those listed above and copper) has been made (Wharfe and van den Broek, 1977) in both fishes and the larger invertebrates. These authors report that all levels were low and mercury residues were comparable with reported values elsewhere, with the exception of the liver of an eel of a size greater than 300 mm (12 in.) which was somewhat higher. Lead and cadmium were generally higher in fish livers than in muscle, but neither was much higher than has been reported in the literature. Not surprisingly, the highest levels were found in animals close to industrial complexes. These results are not necessaiily applicable to the Thames, because of the totally dissimilar physical configuration of the two estuaries and because residues of this kind depend on local industries and their concentration in the estuary, but with this reservation it does appear to be unlikely that Thames sea fish contain unacceptable levels of these toxic metals. However, studies of heavy metals in fishes and invertebrates in the Severn estuary (Hardisty *et al.*, 1974) have shown how the levels of cadmium and other metals vary from species to species, probably depending on their diet, and also show interesting variations such as the high levels of zinc found in brook lampreys from unpolluted inland rivers which were higher than those in the related river lamprey in the polluted regions of the Severn. If anything, these results show that caution is needed in interpreting this kind of information, and means that continuing studies of dangerous contaminants of fish and invertebrates in the Thames will be desirable.

If concentrations of heavy metals were to be found they would most likely have occurred in the region of the sewage sludge dumping grounds in the Black and Barrow Deeps. The dumping of sludge, which is the solid element retained in the sewage treatment works, was begun in 1887, and until 1915 it was discharged into the Barrow Deep. From 1915 to 1967 it was dumped in the Black Deep, and thereafter in the Barrow Deep again. To determine whether any damage to the fauna was caused, or if there was build-up of toxic substances, Shelton (1971) made a study of these two deeps. His conclusions were that the dumping of sludge caused very little effect on the dissolved oxygen in the water even on the bottom and that the fauna showed no evidence of diminution when compared with that of adjacent channels (in which sludge had never been dumped). While the invertebrate fauna was only moderately rich this was probably due to the strong tidal currents in the outer estuary and the high suspended-solid loads of the water of the estuary. Interestingly, his results suggest that for nearly a century London's sludge has been dumped in the outer estuary with no serious effects. This is probably due to the strong scour of the tides in the area which has redistributed the sludge into the North Sea. It is in striking contrast to the situation off New York where large areas of completely anaerobic sediments with an overlying layer of water of low oxygen content have been found.

The relationship between the temperature of the river water and dissolved oxygen has already been touched on. Higher than normal temperatures may also affect the migration of salmonid fishes, especially the downstream migration of the young fish or smolts. The temperature of the river thus has a double significance today when it is inhabited by fish, than in the period during the 1950s when no aquatic life was present. In the years of 1920–61 the temperature of the water had risen by an average of 0·09°C annually (that is 1°C every eleven years) a factor which in the later years was naturally affecting the level of dissolved oxygen. Most of this heat was discharged by electricity power stations beside the Thames which used direct cooling by river water. During the 1960s as some of the upstream stations were taken out of commission the river temperature began to decline. The critical nature of the river temperature makes future plans for new power stations of particular interest. Fortunately, for a variety of reasons the proposals for new stations are downstream of central London. New stations are planned for Brunswick Wharf (capacity 900 megawatts), Barking (capacity 2,000 megawatts), and Littlebrook (capacity 2,000 megawatts), and all are expected to be commissioned by the late 1980s. The effects of the two large-capacity stations are likely to be considerable, but the total rise in temperature in low-flow conditions of the second quarter of the year is likely to be limited to about 3°C above the estimated temperature of 1980. This in turn might depress the dissolved oxygen over 50 km (31 miles) by up to 5 per cent.

It is clear that these proposed power stations will make a considerable, but possibly not serious, impact on the lower tidal Thames and its fishes. Its effect on the possible re-establishment of salmon in the river is summed up by the Thames Migratory Fish Committee's Report (Thames Water Authority, 1977) as follows:

> The conditions for the passage of smolts might be judged to be marginal and mortalities might occur in some years . . . . Similarly, the predicted levels of dissolved oxygen in the second and third quarters may be considered to be marginal for the passage of adult salmon and sea-trout, but the conditions in the fourth quarter could well allow their passage, especially since the motivation to migrate may increase as the spawning season approaches.

Clearly, any future developments in the long term which might affect the temperature regime of the river will need to be scrutinised very carefully with their impact on the ecology of the river in mind.

The elevated temperature of the Thames has a bearing on the use of the Thames barrier now nearing completion at Woolwich. This barrier is essentially for flood prevention, for it is a sad fact that London (and the south-east of England generally) is sinking at a rate of approximately twelve inches a century. This means that large parts of the city would be vulnerable to flooding should the equinoctial spring tides coincide with a 'tidal surge' in the North Sea. The tidal surge is a result of strong winds from the north following a trough of low pressure moving in from the Atlantic pushing water down into the southern North Sea, and the Thames estuary in particular, which results in high tide being much higher than normal. This happened in 1953 when large parts of the east coast (and the Netherlands) were flooded. A result of the sinking of the land is that the comparative height of the tide has increased, about 300 mm (12 in.) at Southend, and nearly 600 mm (23·6 in.) at London Bridge, with very real danger of disastrous flooding in London. Although in the past the tide has been kept at bay by raising the river wall, the wall is now at a height in many places where the average passer-by cannot see the river because of it, and further heightening of it, which would of necessity have to include docks, tributaries, and other embankments, is no longer feasible. The Greater London Council, as an alternative, embarked upon a most ambitious and far-sighted plan to construct a barrier at Woolwich which, when flooding threatened, could simply be raised out of the river bed to hold back the tidal water. This plan, however, required the river wall downstream of the barrier and in other places where the flood water might spill over onto the land to be raised.

The plan for the barrier is that it will normally lie in the river bed, so as to allow shipping to pass upstream, and will only be raised when flood conditions threaten (except for the occasions when practice is required). This may be only once or twice a year in the early years after its construction but by the end of the century it may need to be raised up to ten times annually. As envisaged its use poses no threat to the wildlife of the Thames. Unfortunately, there is a body of opinion that would like to see the barrier permanently raised, and once it is completed the demands of this faction may well increase. It is appropriate therefore to put on record the ecological disadvantages of doing this. The upstream section of the tideway would be ponded back; a substantial part of this water is treated sewage effluent (albeit treated to a high standard) and in addition several large power stations are upstream of the barrier (including the new 900 megawatt station at Brunswick Wharf) so the heat gain to the river would be considerable and the quality of the water may be poor at times. Another physical disadvantage that would follow is that large quantities of silt would fall out of suspension in the water, and the immediately upstream section of the barrier would collect large volumes of mud which would need to be dredged out. Most important of all if the barrier were used in this way it would represent an insurmountable obstacle to migratory fishes, especially the smelt, flounder, sea trout, and salmon (if it is ever re-established), although elvers might very well find their way around it. The consequences of the proposed action by this small body of opinion would thus be disastrous to the fish fauna and aquatic invertebrates.

Another threat to fish life, and one which would become more serious if the barrier was kept permanently in position is the discharge of storm water into the river. This storm water, as its name implies, is water resulting from a heavy rainstorm which reaches the sewers through drainage from gutters and roadside drains overfilling the sewers. Normal light rainfall presents no problem as the sewers have the extra capacity to take this besides the usual sewage flow, but in freak conditions of heavy rainfall the sewers overflow into storm water channels which discharge into the Thames, carrying mostly rainwater but also untreated raw sewage. Some of these discharges are pumped out and the volume of storm water discharged is known but even as late as 1973 one authority was unable to predict the effect of such discharges because some of them, the so-called gravity discharges, which at times could be very large, were neither metered nor sampled. In simple terms no one knew exactly how much untreated sewage was reaching the Thames in such conditions which locally might happen once or twice a year. The storm water removes the dissolved oxygen from the river water and as a result the fish may be distressed.

That storm water discharges could have a serious effect on freshwater fishes in the river was shown in June 1973 when around Battersea Bridge and in Wandsworth Reach a large number of fish, probably thousands, were killed following torrential rain and consequent storm water discharges. Another fish kill under similar circumstances was observed in the Woolwich region in August 1977 when chiefly smelts were killed. The unfortunate fact is that when the river was grossly polluted and fish could not inhabit it these discharges passed unnoticed, they simply made a bad state a little worse, but once fishes were established in the cleaner

Thames the effects of putting storm water into the river became noticeable. It must be admitted that alleviation of this threat to the fish and aquatic wildlife is not simple and its cure would be expensive. However, in the present context of concern for the environment this is a situation which cannot be tolerated. Some means of preventing the untreated sewage carried by storm water from reaching the river will have to be devised even if it is the expedient of constructing additional drains for surface water and storage tanks in which the storm water can be held until it can be treated.

While freak rain storms pose problems of one kind the opposite meteorological hazard, drought, may produce changes in the fauna, and possibly deterioration in quality of fish life. The extensive drought of 1976, when in the Thames catchment area for the twelve-month period prior to September less rain fell than in any year since record-keeping began in 1727, produced a massive intrusion of salt water into London together with a near marine fauna. At high tide on occasions this salt water penetrated upstream as far as Eel Pie Island, only 1 km (0·6 miles) below Teddington Weir (Andrews, 1977). While it is of some interest to record the capture of sprats in Chelsea Reach and prawns at Chiswick, as Mr Andrews did during that year, the condition of the freshwater fishes which were 'trapped' between the salt water and the weir at Teddington can only be imagined. It is probably as a consequence of this that anglers in that area suffered one of the poorest seasons in terms of both numbers of fish caught and average weight of specimens of each species that they had experienced for a decade (see p. 137), a trend that was even more noticeable in the following year.

There is no doubt that the conditions in this area were worsened by the so-called 'tail-to-head' pumping at Teddington lock from July to September, which 'replaced' water lost through the navigational locks or by leakage through the weir structures. This enabled the maximum amount of water to be abstracted from the river to supply London's drinking water. During this drought conditions were so extreme that it was only by the utmost and most creditable efforts of the Thames Water Authority's staff that drinking water supply was maintained. In the tideway one result of this tail-to-head pumping, where water was pumped from below the weir to above it, was a severe deterioration in the habitat occupied by the freshwater fish. It is perhaps fortunate that such climatic conditions have occurred only once in the last two-and-a-half centuries, but the possibility of more frequent occurrence in the future due to climatic change cannot be ruled out. What is more certain, with the rising demand for domestic water in London and the south-east of England, together with rising sea-level relative to the river bed, is that in the future droughts of much less severity may produce the same effect. There are thus grounds for some concern for the future quality of life in the upstream reaches of the tideway which add further urgency to the solution of the region's water supply problem.

FACTORS AFFECTING FISHES IN THE TIDEWAY

While the Thames now contains a varied and at times abundant population of fishes there are still a number of factors which influence the size and composition of this fauna. As has been pointed out in chapter 6, in the period at which recolonisation was taking place there was in general a marked scarcity of young freshwater fishes caught in the metropolitan reaches of the river by power stations. This was probably a true reflection of the population structure of the species because young fishes, being weaker swimmers, are more commonly caught by power stations than are large ones (as is shown by the present situation at West Thurrock power station, where sometimes thousands of young whiting and flatfishes are caught during a tidal cycle). So far as the freshwater fishes, such as roach, bream, dace, and perch, were concerned, the virtual absence of young fish suggested that they were not breeding in the metropolitan reaches of the river, but were recruited at an age of one year or more from upstream. Upstream of London, and especially in the Richmond to Teddington area netting captured numerous young fish, and other observations confirmed that in these reaches the freshwater species were breeding.

This situation highlighted one of the problems of the freshwater reaches of the tideway, a scarcity of suitable habitats for spawning. This had little to do with the pollution history of the river but was more a reflection of the controlled nature of the river banks and flow. Oxbow lakes and backwaters, pools and still waters on water meadows, such as one finds in more natural rivers and which form eminently suitable spawning areas and nursery grounds, have been excluded from the flood plain of the tideway. As a result, the only areas where spawning is possible are in a few sheltered areas in the lee of the eyots or bends in the river, where aquatic vegetation can grow. These are very restricted below Richmond. Above the lock at Richmond there are more suitable areas, and because the water is pounded back by the lock the river bed is not subject to the strong tidal scour which keeps so much of the river bed bare in the tideway proper. It seems very likely, although investigations into this topic have not been made, that very few of the fish in the tideway have actually been bred there but have moved down from the only semi-tidal reaches between Richmond and Teddington. There is no evidence that significant numbers of fish have entered this region from the non-tidal Thames above Teddington.

As a corollary of this it could be suggested that the conservation measure most needed in the upper tideway is the provision of suitable spawning habitats. Because of the tidal scour the main river is unsuitable for such a measure. Ideally, disused docks in which water plants could grow and provide the necessary places for attachment for the eggs of plant-spawning cyprinids, perch and pike, would prove the simplest and cheapest means of replenishing the freshwater fish stocks of the tideway. Such a dock could be adapted by the construction of a permanent sill across its mouth so that communication with the river was maintained without the full tidal current or rise and fall affecting the habitat adversely. With the economic and social changes in upstream docks that have been taking place in the London aıea, suitable docks could well be available. While such an adaptation is advanced here mainly as a suggestion to provide a suitable habitat for freshwater fish to breed in, it clearly would serve other purposes, such as 'quiet area' within the metropolis for rest or recreation, it would also probably attract water birds, thus enhancing the amenity to naturalists, and would provide a suitable still water for educational visits and study (in an area which is poor in such features).

Study of the growth rates of the freshwater species such as roach and dace showed that in the tideway they grew well in their first years but thereafter growth was poor. The suggestion advanced to explain this was that there was known to be an abundance of small crustaceans in the near-surface waters of the river which was suitable as food for young fish, but that the larger organisms that bigger individuals required were in most of the river scarce. Accordingly, growth was limited by shortage of food once a fish had attained a certain critical size. It will be interesting to see whether this is just a phenomenon associated with the early stages of the re-establishment of a natural fauna (for my observations were mostly based on the period 1967–73); possibly with time a richer invertebrate fauna will be established which will in turn lead to better growth amongst the larger fishes. However, in view of the tidal nature of the river, its narrow course through the metropolitan reaches, and the unpromising nature of its muddy bottom through much of its length, it may be that food for the larger, invertebrate-feeding species will always be scarce and their growth slow.

Although the number of bream examined for age and growth studies was small, this seemed to be one species which grew well throughout life. I attributed this difference to the abundance of tubifex worms in the river during the period 1967–73. The small blood-red worms were abundant in the river even when it was heavily polluted and were subject of a small 'fishery' in which the Port of London Authority licensed collectors to harvest the crop for sale to aquarium shops and dealers (tubifex worms being an important live food for aquarium fishes). Study of stomach contents, and the presence of parasites which are carried only by tubifex worms, showed that the bream must have fed heavily on them, and due to their abundance they must have supplied adequate nourishment for both large and small bream, for there was no sign of a dramatic decrease in growth after the first few years of life. Harrison and Grant (1976), in their fascinating account of the waterfowl of the Thames, report that some birds, especially the dunlin, were also exploiting tubifex as an important food source. They also have shown that with the change in the quality of the water the numbers and species of tubifex worms have altered. In highly polluted water *Tubifex tubifex* and *T. costatus* thrive; *Limnodrilus hoffmeisteri* requires water of a higher quality, and another species, *Peloscolex benedeni*, cleaner water still. These four species have different salinity preferences also; *T. tubifex* lives in freshwater as does *L. hoffmeisteri*, although the latter can tolerate low salinities; *T. costatus* can live in a wide range of salinities and *P. benedeni* prefers near-marine conditions. Not surprisingly, with the dramatic decrease in pollution in the Thames those species adapted to live in polluted water have become rarer. Figures quoted by Harrison and Grant for the London Bridge area show that the density of tubifex worms in 1971 was 300,000 per square metre and in 1975 4,000 per square metre. Many of the former tubifex fishermen of the Thames foreshore have found that it is now no longer possible to collect sufficient worms to supply the pet-fish shops. In addition, the composition of the tubifex worm fauna in the Thames has been altered by the greater penetration of salt water upstream during the relatively dry years of the middle 1970s. It is said by Harrison and Grant that these changes in the tubifex populations have affected the numbers and distribution of wildfowl on the Thames. Most particularly birds such as pochard and tufted duck, which feed heavily on the worms, have declined appreciably in recent winters on the 'Inner Thames'. It seems that much of the phenomenal increase during 1968–75 in water birds in the London Thames may have been due to the decrease in pollution making the water less noxious to the birds and exposing an unexploited food resource in the shape of tubificid worms. As these declined as the water became cleaner locally the birds tended to occur further upstream, and now with the diminution of the worm population the bird revival has proved to be largely ephemeral.

How the changes in tubifex populations and numbers will affect fish in the river is not known. It may be that other invertebrates will increase in numbers and provide an alternative food source for bottom-feeding fish like the bream. Other fishes would be capable of eating these invertebrates as well, so that the overall view may be that the bream will find the Thames in London a poorer place to live but that there will be more food available to the larger members of the freshwater fish community as a whole.

In the period 1968–73 when regular monitoring at several power stations along the Thames was undertaken there was a distinct seasonal variation in the numbers of fish. Throughout the winter numbers of both freshwater and marine species were high, while in the third quarter of the year, July–September, they were low. This was especially noticeable in the years 1968–71 at the stations in the Outfalls and the Dartford Tunnel areas when not infrequently no fish were caught for four or five months. However, this began to change in 1972 and small numbers of fish were caught throughout the year in these areas and were certainly present at the upstream stations. Three factors were involved in this decrease in the summer months, one of which had no bearing on the number of fish present. In England demand for electricity is at its lowest at this period and power stations often work to less than full capacity or may be shut down for repairs and alteration during this 'slack' time. It may be that this affected the total catches of fish, as some Thames-side stations were on less than full production and consequently using less river water. Although possibly significant this does not fully account for the decline in numbers of fish caught.

A more important factor so far as fishes are concerned was that during the third quarter period water temperatures are normally high and freshwater flow low and as a consequence the dissolved oxygen in the water is at its lowest level. It seems probable that the low levels of dissolved oxygen deterred freshwater fishes from moving downstream, while those already in the depleted area moved upstream as conditions approached the level of distress. By 1972 it seems that the third-quarter conditions were less adverse and from that year onwards fishes began to inhabit or occur in the middle reaches of the tideway throughout the year in some numbers. The occurrence of sea fish is, however, also affected by the migrations made by each species during the year, and the general inshore migration of young fish mainly at the end of the third quarter heightened the contrast with mid-summer when no fish were being captured at the downstream power stations. Sea fishes, like the freshwater species, were scarce or apparently absent in summer until 1972, but from the next year they began to occur in increasing numbers. Today, some species, such as the whiting, sole and flounder, are found in some numbers throughout the summer months.

The dominant factor in the distribution of fishes in the tideway seems no longer to be the level of dissolved oxygen (except locally after storm water discharges) but the salinity of the water. The drought of 1976 showed that marine fishes, such as sprats and gobies, had penetrated upstream into the heart of London and it is a fair assumption that the freshwater fishes were forced upriver into the upper part of the tideway, with serious consequences to their condition and numbers. With more normal freshwater flows neither species would be likely to occur further upstream than the Blackwall Tunnel area, while in conditions of heavy freshwater flow they would be further downstream still. That this kind of colonisation occurs within a year adds emphasis to the statement that in general fish are highly adaptable animals and will take advantage of suitable living space that becomes available. This is even better illustrated by the return of fish to the tidal Thames as a whole.

Sea fishes are also especially subject to the effects of tides and storms. While the abundance of haddock in the Thames in 1970 may have been ultimately due to the exceptionally strong year classes producing a wealth of fish in the North Sea, their actual appearance in the tideway occurred shortly after strong north-easterly winds had blown for several days. It can be inferred from this that the wind had produced a movement of North Sea water (with its contained organisms) into the mouth of the Thames. Storms may account for the presence in the mouth of the river of a number of species which are normally found in the North Sea, or the eastern English Channel, as for example, the poor cod, ballan wrasse, and marbled electric ray. In addition, the variation in the tidal cycle is also observable in the abundance of such species as the sole, which appears to increase markedly in spring and autumn following the equinoctial spring tides. As it is known that the sole has a well-developed capacity for using tidal currents for transport it is a reasonable assumption that the Thames soles are being carried by these highest tides of the year. It is also noticeable that the main influx of fish such as whiting, sand gobies, sprats and herrings begins with the autumn equinoctial spring tides.

FUTURE DEVELOPMENTS AFFECTING FISH LIFE

The improvement in the Thames as a wildlife habitat has produced several tantalising situations. While conservationists and the general public alike welcome the return of water birds and fishes to the river, industrial users of Thames water have discovered that their intake and other pipes have become the homes of animals which reduce water flow or block the plumbing entirely. Huge quantities of mussels, whiteweed (which looks like a plant but is actually a colony of animals), barnacles, and sea squirts may attach themselves to intake culverts, periodicaly breaking loose to block up some vital pipe. Wharves made with timber piling are feared to be at risk from attacks by the timber-eating ship-worm, *Teredo*, and certainly will attract other wood-boring animals in the sea reaches. The Thames has even been 'invaded' by the Chinese mitten crab, *Eriocheir sinensis*: three specimens were taken at West Thurrock power station in February, May and June 1976, and a fourth was caught at Teddington in September 1977, the specimens probably being transported across the North Sea from the River Elbe, where it is common (and to which it was introduced from China in 1912). This crab makes burrows in river banks and has been said to cause considerable damage in Europe, as well as altering the biological composition of the river.

Some, if not all of these animals living in the Thames may prove to be unwelcome to users of the water or industry on the river bank, and it is a paradox that in the days of gross pollution such animals could not live in the greater part of the river and so could not become a problem. The presence of a rich aquatic fauna will also mean that planned development or alteration to the river in any large way will now be scrutinised with special vigilance in case the changes have any adverse effect on the fauna or the balance of life in the river. Every development must now be considered in the light of a living river. One such, which was discussed by Harrison and Grant (1976), is the construction of a new waterfront at Thamesmead on the Erith Marshes, downstream of Woolwich, which involves the 'reclaiming' of the foreshore by the construction of a new sea wall. As these distinguished authors point out, this single development will destroy a feeding area which a short time before was frequented by up to 7,000 wild aquatic birds. It must, in fairness, be admitted that the plans for this destruction of the habitat were made before its importance to wild fowl had been recognised.

The recognition that the mouth of the river and the inner estuary o the Thames is an important nursery ground for young fish, especially the commercially valuable sole, plaice, and whiting, has added a new dimension to plans for industrial development in this region. Petro-chemical works, and even oil storage along the lower tideway are an ever-present threat to these young fish. More serious is the much discussed deep-water port off Foulness, the island which, having once escaped the environmental disaster of an international airport, is now again threatened with development of its southern side as a port. If this plan by the Port of London Authority merits further discussion it should be considered in the light of the consequent destruction of a large part, and major alteration of the remainder, of these nursery grounds. It must be realised that these fish are not just Thames-mouth fish; after their nursery stage is past they contribute to the stock of the southern North Sea. So the choice is not a simple one of a new port versus a small supply of fish locally, but a new port versus damage to the stock of fish in the southern North Sea.

The Thames barrier, now nearing completion at Woolwich, has already been mentioned. If used as its original plans and policy intended, it is an essential and imaginative way of safeguarding London from flooding, although in the long term one wonders how the sea can be kept out for ever. However, that redoubtable propagandist for a tideless Thames, the late Sir Alan Herbert, had bequeathed the proposition that a permanent barrier would be desirable and this is still advanced by a vociferous lobby within London. As already pointed out, the use of the barrier as a permanent barrage will cause great damage to the aquatic fauna of the river and most of all to the migratory fishes the return of which in the present decade stands as the most remarkable product of the cleaner Thames.

These few examples go to illustrate the problems that may face the tideway and its wildlife in the future. Development of any kind in the river or the estuary, has now to be scrutinised for its effect on the wildlife, as well as amenity or visual impact.

THE VALUE OF FISH IN THE TIDAL THAMES

In the initial stages of the great 'clean-up' of the tidal river the presence of fish life, where once there had been none, represented a demonstration of success which the layman could appreciate. No matter how many times the chemists announced that dissolved oxygen was present in the water all year round the resulting interest was minimal. But a fish living in a river was something tangible, whereas dissolved oxygen was tasteless, invisible, and impossible to visualise. At the outset, then, the fish returning to the river presented a visible example that the press and other media could use, culminating in 1974 when a live salmon was caught at West Thurrock. Today, fish in the river offer more than a useful public relations point to be discussed whenever the subject of water pollution in London is raised, although there is still considerable interest in the success story of the Thames.

As we have seen, the mouth of the Thames and the estuary as a whole is an important nursery ground for fish, many of them species of considerable commercial importance. Stock from the estuary moves out into the southern North Sea on maturity and supplies the important commercial fisheries in the area. The importance of the estuary nursery grounds cannot be over-emphasised, not least as they are entirely within British fishery limits and therefore fishing can be strictly controlled. In addition to the commercial species which live in the Thames when young there are species, such as sprats, which are in sufficient abundance to be exploited within the mouth of the river. Shrimps too are now abundant enough to form fisheries within the river. However, due to social and economic changes the Thames-mouth fisheries will probably never be so important as they were in the seventeenth and eighteenth centuries when freshness of Thames fish was a major advantage over fish from the west coast or the Channel. The major prospects for the Thames fishery probably lies in the development of angling. The point has already been made (p. 176) that early in the 1970s members of a Gravesend angling club found that they could catch more fish off their own Thames foreshore than during an expensive and time-consuming outing to the Channel coast. This is the great advantage of the Thames, lying as it does beside so many centres of population from London downstream to Sheerness and Southend, and with road and rail communications generally following its course. Anglers can now expect to catch worthwhile fish in the lower reaches, and although the number of species is restricted in comparison with those in the western Channel or the Atlantic coast, there is a wide variety available. Further out into the estuary, the fishing for roker, smooth-hound, cod (in winter), grey mullet, bass, plaice and sole has always been reasonable and may now be better because these species are likely to move more frequently into the Thames mouth.

Upstream in the tideway freshwater fishing has always been possible and at times good above Chiswick, although it is vulnerable to a number of threats, chiefly to reduced freshwater flow from the non-tidal river and also local pollution. However, in normal circumstances freshwater fishing is possible with reasonable chances of success downstream at least to the Westminster Bridge area, and for some species below this. The 'fishing experiments' organised by the Greater London Council and the Thames Angling Preservation Society proved that fish could be caught by anglers in the heart of London. Subsequently anglers have discovered that smelt, rainbow trout, eels, and flounders can be caught in the reaches between London Bridge and the Blackwall Tunnel, and that disused docks, like the Shadwell Basin, now offer considerable potential for angling. The importance of this fishing is not the size or numbers of fish caught (specialist anglers who require large catches or record sized fish will know that they have to travel to achieve their objects), but that the angling lies through the heart of a densely populated city in which opportunities for open air recreation are limited, and in which travel is expensive. The opportunities offered for recreational angling are thus especially valuable, for example, a rainbow trout caught in 1977 on the foreshore at Greenwich by a local schoolboy; that the river can offer fishing of any kind is a major advantage in an area where angling waters are scarce.

The re-establishment of salmon in the Thames is being actively pursued by the Thames Water Authority, following the report of the Thames Migratory Fish Committee which was originally set up by the Port of London Authority. This committee, after considering predictions of the future dissolved oxygen and temperature regimes of the estuary by 1980, considered that these factors would not be an obstacle to migrating salmon or sea trout and that there was a reasonable chance of success for an experimental re-introduction of both. However, the problem in restoring a natural population of these fish (i.e. migrating into the river, spawning, and migrating downstream unaided) lies in the limited natural spawning grounds that the river now offers and the numerous weirs and locks on the river above Teddington (many of which would be impassable in conditions of normal river flow). Initially, the proposals suggest that parr of both species from rivers in southern England should be released in limited areas, the fish marked so that their survival rate, growth, and movements can be followed by later surveys. Proof that these fish had migrated to sea and returned would be obtained when adult fish were captured at a trap to be installed at Molesey weir (Teddington weir having been modified to make it less of an obstacle to the returning adults). Unless all the weirs (and there are forty-four on the main river between Lechlade and Teddington) were made passable by the construction of fish passes, which would be immensely expensive, the only way of getting adult fish to the spawning grounds would be to trap them at Molesey weir and transport them overland to the headwaters or raise their progeny in a fish hatchery. Clearly, not until several generations of fish had made the complete cycle from Thames hatchery or nursery ground to sea and back to the Molesey fish trap, and the numbers of fish had built up significantly, could the pilot scheme be considered successful and the further heavy expenditure involved in adapting weirs to allow fish to migrate be justified.

The attractions inherent in re-establishing salmon in the Thames are clear. There have been several such attempts already (by the Thames Angling Preservation Society in 1862–4, and by the Thames Salmon Association in 1901–6) all of which failed, and more recently salmon parr have been released in the river in well-publicised but ill-conceived small-scale exercises. The real problem, however, is a human one, not necessarily involving physical or chemical barriers. In general, the anglers of the Thames, both on the tidal and non-tidal reaches, seem largely indifferent to the prospect of having salmon in the river. If any opinion is expressed it is usually opposition because they fear that the rent they pay for exclusive rights to fish a stretch of bank may be raised as a result of potential salmon fishing. In practice this may not happen, but nevertheless it is a view held by several well-known angling administrators and writers. In addition, there is a very real possibility that if salmon were to be re-introduced at considerable expense they might well be captured in the mouth of the Thames (outside the Thames Water Authority area) by the commercial fishermen who take large numbers of bass and grey mullet (together with occasional salmonids) in surface-fishing nets. Clearly, it would be possible to legislate to prohibit or control the use of nets for the intentional capture of salmon, but the nets at present used for the intentional capture of bass and other surface-swimming fish are precisely those that would capture salmon.

Any proposal to expend considerable sums of money in the provision of fish hatcheries and fish traps solely for the purpose of re-establishing salmon in the Thames should be subject to close public scrutiny, even more so if the proposals include provision for building fish passes into the existing weirs along the river to allow the fish to migrate, which could increase the expenditure by millions of pounds. Comparisons with the millions spent to alleviate pollution in the tideway are not valid, as this expenditure was incurred in the interests of public health and to remove the nuisance caused by the polluted water in the metropolis – the return of fish and other wildlife was simply a welcome sequel. Re-introduction of salmon, however, would be an expensive operation, wanted by very few people and with the sole object of providing salmon and sea trout fishing in the Thames (possibly a case can be argued that it is required of the Authority under the Salmon and Freshwater Fisheries Act of 1975 to maintain, improve and develop all fisheries). Although obviously success would be the public relations exercise of the century, any such expenditure should be questioned in the light of the other outstanding problems of the tideway and its fishery. An effective means of dealing with the storm-water discharges into the river within the metropolitan area which seriously affect the migratory and freshwater fish after heavy rainfall needs to be found. The intensive abstraction of water from the non-tidal Thames and many Thames tributaries has seriously affected the quality of the habitats in the headwaters especially; and there is no worse example than the headwaters of the Lee, which in forty years have virtually been extinguished. The dramatic drought of 1976 produced a desperate worsening of the freshwater habitats of the upper tideway, enhanced by the 'tail-to-head' pumping at Teddington weir. Although this was an exceptionally severe drought and abstraction of virtually the whole Thames flow was necessary to supply drinking water to London, the situation demonstrates how critical the water supply situation is in the lower Thames basin. Both storm-water discharges and heavy abstraction of freshwater could pose serious threats to salmon and sea trout in the lower Thames. A more realistic scale of priorities would therefore seem to be to solve these problems before attempting even an experimental restocking of the Thames with salmon and sea trout. In the meantime the tideway offers moderately good angling, and has a healthy stock of non-salmonid fish with occasional brown trout and sea trout (which will naturally become more numerous as time progresses). No doubt the fish stocks could be improved, as could those of the non-tidal river upstream of Teddington, but to commit slender resources to the re-introduction of salmon is as unnecessary as it is irrelevant to the needs of anglers, and to the continued improvement of the river as a wildlife habitat.

REFERENCES

ANDREWS, M., 1977, 'Observations on the fauna of the metropolitan River Thames during the drought in 1976', *London Naturalist* 56: 44–56.

BARRON, R. J. C., 1976, 'The occurrence of the rocky goby, *Gobius paganellus* L. 1758 and the two-spot goby, *Chaparrudo flavescens* (Fabricius, 1779) in the Blackwater Estuary, Essex (S.E. England)', *Journal of Fish Biology*, 8: 93–5.

BATES, L. M., 1977, 'A river fit for fish', *Wildlife*, 19: 35–8.

BELYANINA, T. N., 1969, 'Synopsis of biological data on smelt *Osmerus eperlanus* (Linnaeus) 1758', *Fishery Synopsis 78*, FAO, Rome.

BINNELL, R. (ed.), (by R. Griffiths), 1758, *A Description of the River Thames*, . . . T. Longman, London.

BROUGHAM, W. H., 1893, *Rise and Progress of Thames Preservation*, Sampson Low, Marston, London.

BUCKLAND, F. T., 1863, *Fish Hatching*, Tinsley, London.

BUCKLAND, F. T., 1879, *18th Annual Report of the Inspectors of Salmon Fisheries*, HMSO, London.

BUCKLAND, F. T., 1883, *Curiosities of Natural History* (2nd series), R. Bentley, London.

BUCKLAND, F. T. and WALPOLE, S., 1879, *Report on the Sea Fisheries of England and Wales*, HMSO, London.

BURGESS, G. H. O., 1967, *The Curious World of Frank Buckland*, J. Baker, London.

BURRETT, J., 1960, *Fishing the Lower Thames* (2nd edn, 1968, *Freshwater Fishing : the Lower Thames*), Benn, London.

[CAMERON, A.], 1970, 'The Dutch eel craft in London', *Port of London*, 45 (538): 206–10.

CARGILL, C., 1969, *The River Thames – Historical Survey of the Rights over the River and their Conservancy*, Thames Angling Preservation Society, London.

CARGILL, C., 1972, *The Thames Angling Preservation Society, Past, Present and Future*, Thames Angling Preservation Society, London.
CHANDLER, D. and LACEY, A. D., 1949, *The Rise of the Gas Industry in Britain*, British Gas Council, London.
CHAUNCEY, SIR HENRY, 1700, *The Historical Antiquities of Hertfordshire*, Ben Griffin, London.
CLARIDGE, P. N. and GARDNER, D. C., 1977, 'The biology of the northern rockling, *Ciliata septentrionalis*, in the Severn Estuary and Bristol Channel', *Journal of the Marine Biology Association of the UK*, 57: 839–48.
CLARKSON, L. A., 1971, *The Pre-industrial Economy in England 1500–1750*, Batsford, London.
COHEN, I., 1955, 'Apprentices and salmon', *Transactions of the Woolhope Naturalists' Field Club*, 35 (1): 7–18.
CORNISH, T., 1902, *Naturalist on the Thames*, Seeley, London.
CRAGG-HINE, D., 1969, 'The feeding habits of fish in the effluent channel of Peterborough power station', CERL Report No. RD/L/R 1556: 1–7.
CRAGG-HINE, D., 1970, 'The reproductive cycle of fishes in the effluent channel of Peterborough power station', CERL Note No. RD/L/N 157/69: 1–7.
CROFTS, J., 1967, *Packhorse, Wagon and Post*, Routledge & Kegan Paul, London.
DAY, F., 1887, *British and Irish Salmonidae*, Williams & Norgate, London.
DEELDER, C. L., 1970, 'Synopsis of biological data on the eel *Anguilla anguilla* (Linnaeus) 1758', *Fisheries Synopsis 80*, FAO, Rome.
DEFOE, D., 1724, *A Tour thro' the Whole Island of Great Britain* . . . , 3 vols, G. Strahan, London.
DOCKLANDS JOINT COMMITTEE, 1975, *Conservation and the Role of the River. Working Paper 8*, Docklands Joint Committee, London.
DONOVAN, E., 1802–8, *Natural History of British Fishes* . . . , 5 vols, London (no publisher).
EMMISON, F. G., 1976, *Elizabethan life : Home, Work and Land from Essex Wills and Sessions and Manorial Records*, Essex County Council, Chelmsford.
FARMER, J., 1735, *The History of the Ancient Town and Once Famous Abbey of Waltham . . . Essex*, London.
Field, 1901, Editorial note in issue for 28 December 1901, no. 2557, p. 1004.
GAMESON, A. L. H. and WHEELER, A., 1977, 'Restoration and Recovery of the Thames Estuary', in J. Cairns *et al.*, *Recovery and Restoration of Damaged Ecosystems*, University Press of Virginia, Charlottesville, pp. 72–101.
HARDISTY, M. W. and HUGGINS, R. J., 1975, 'A survey of the fish population of the middle Severn estuary based on power station sampling', *International Journal of Environmental Studies*, 7: 227–42.
HARDISTY, M. W., HUGGINS, R. J., KARTAR, S., and SAINSBURY, M., 1974, 'Ecological implications of heavy metal in fish from the Severn estuary', *Marine Pollution Bulletin*, 5: 12–15.

HARRIS, M. T. and WHEELER, A., 1974, '*Ligula* infestation of bleak *Alburnus alburnus* (L.) in the tidal Thames', *Journal of Fish Biology*, 6: 181–8.

HARRISON, J. and GRANT, P., 1976, *The Thames Transformed : London's River and its Waterfowl*, Deutsch, London.

HARTING, J. E., 1894, 'Isaak Walton's Association with the River Lea. With some notes on the former existence of Salmon in that River', *Essex Naturalist*, 8: 186–98.

HERBERT, A. P., 1966, *The Thames*, Weidenfeld & Nicolson, London.

HIBBERT, C., 1969, *London. The Biography of a City*, Longmans, London.

HOBSON, J. M., 1924, *The Book of the Wandle : The Story of a Surrey River*. Routledge, London.

HODGSON, W. C., 1957, *The Herring and its Fishery*, Routledge & Kegan Paul, London.

HORN, P. W., 1923, 'Notes on the fishes of the London Docks', *London Naturalist*, 1922: 19–21.

HOWARD, P., 1975, *London's River*, Hamish Hamilton, London.

HOWSON, J., 1975, *A Brief History of Barking and Dagenham*, London Borough of Barking, Barking.

HUDDART, R. and ARTHUR, D. R., 1971, 'Lampreys and teleost fish, other than whitebait, in the polluted Thames estuary', *International Journal of Environmental Studies*, 2: 143–52.

JESSE, E., 1834, *Gleanings in Natural History* (2nd series), J. Murray, London.

JONES, J. R. E., 1964, *Fish and River Pollution*, Butterworths, London.

KOOIJMANS, L. P. L., 1972, 'Mesolithic bone and Antler implements from the North Sea and from the Netherlands', *Bericht van de Rijksdienst voor het Oudheidkundig Bodemonderzoek*, 20–1: 27–73.

KORRINGA, P., 1967, 'Estuarine Fisheries in Europe as Affected by Man's Multiple Activities', in Lauff, G. H., *Estuaries*, American Association for the Advancement of Science, Washington, pp. 658–63.

LAVER, H., 1898, 'The mammals, reptiles, and fishes of Essex', *Special Memoirs Essex Field Club*, 3: 1–138.

LONGFORD, E., 1971, *Victoria, R. I.*, Weidenfeld & Nicolson, London.

MACKAY, C., 1840, *The Thames and its Tributaries . . .* , 2 vols, London.

MANN, R. H. K., 1973, 'Observations on the age, growth, reproduction and food of the roach *Rutilus rutilus* (L.) in two rivers in southern England', *Journal of Fish Biology*, 5: 707–36.

MANN, R. H. K., 1974, 'Observations on the age, growth, reproduction and food of the dace, *Leuciscus leuciscus* (L.) in two rivers in southern England', *Journal of Fish Biology*, 6: 237–53.

MANSFIELD, F. A., 1922, *History of Gravesend in the County of Kent*, Gravesend & Dartford Reporter, Gravesend.

MARLBOROUGH, D., 1963, 'A supplement to "The Fishes of the London Area",' *London Naturalist*, 42: 62–70.

MARLBOROUGH, D., 1965, 'London fish since 1962', *London Naturalist*, 44: 70–3.

MARLBOROUGH, D., 1969, 'Further London fish records to 1968', *London Naturalist*, 48: 76–85.

MARLBOROUGH, D., 1972, 'London fishes to 1971', *London Naturalist*, 50: 63–78.

MARSH, R., 1971, *The Conservancy of the River Medway 1881–1969*, Medway Conservancy Board, Rochester.

MEADOWS, B. S., 1971, 'Observations on the return of fishes to a polluted tributary of the River Thames 1964–9', *London Naturalist*, 49: 76–81.

MITCHELL, P. C., 1929, *Centenary History of the Zoological Society of London*, Zoological Society, London.

MURIE, J., 1901, 'Alleged salmon and sea trout in the Thames estuary and adjacent waters', *Field*, no. 2556: 969.

MURIE, J., 1903, *Thames Estuary Sea Fisheries:* I, Kent and Essex Sea Fisheries Committee, London.

MURIE, J. (MS), 'Thames Estuary Sea Fisheries: II', (Fair copy of the manuscript in Reference Section of the Public Library, Southend-on-Sea).

NEWELL, G. E., 1954, 'The marine fauna of Whitstable', *Annals and Magazine of Natural History* (12) 7: 321–50.

PEARSON, A., 1961, *in* BURRETT, J. and PEARSON, A., *Anglers' Angles*, Allen & Unwin, London.

PENNANT, T., 1776, *British Zoology*, vol. 3, B. White, London.

PENNANT, T., 1791, *Some Account of London* . . . (2nd edition), R. Faulder, London.

PIPPARD COMMITTEE, 1961, *Pollution of the Tidal Thames*, HMSO, London.

PORT OF LONDON AUTHORITY, 1967, *The Cleaner Thames 1966*, Port of London Authority, London.

Port of London Authority Monthly, 1933–7, News items in issues dated May 1932, January 1933, and June 1937.

POWELL, W. R. (ed.), 1966, *Victoria History of the Counties of England. A History of Essex*, vol. 5, Oxford University Press, London.

REDEKE, H. C., 1941, *De Visschen van Nederland*, Sijthoff's Uitgeversmaatschappij, Leiden.

ROSE, L., 1975, *Health and Hygiene*, Batsford, London.

RUDÉ, G., 1971, *Hanoverian London 1714–1808*, Secker & Warburg, London.

RUSSELL, F. S., 1976, *The Eggs and Planktonic Stages of British Marine Fishes*, Academic Press, London.

SALMON FISHERIES COMMISSION, 1861, *Report of the Commissioners appointed to inquire into Salmon Fisheries (England and Wales); together with the minutes of evidence*, HMSO, London.

SHELTON, R. G. S., 1971, 'Sludge dumping in the Thames estuary', *Marine Pollution Bulletin*, 1 (2): 24–7.

SHEPPARD, F., 1971, *London 1808–1870: The Infernal Wen*, Secker & Warburg, London.

SMITH, D. J., NICHOLSON, R. A. and MOORE, P. J., 1971, 'Mercury in water of the tidal Thames', *Nature, Lond.* 232: 393–4.

SNOW, J., 1849, *On the Mode of Communication of Cholera*, London.

SOLOMON, D., 1976, 'The decline and reappearance of migratory fish in the tidal Thames, with particular reference to the salmon, *Salmo salar*', *London Naturalist*, 54: 35–7.

TETLOW, J. A., 1971, 'Pollution of the Water', Paper presented at the Twelfth Annual Scientific Meeting of the British Academy of Forensic Sciences, 2 October 1971.

THAMES WATER AUTHORITY, 1977, *Thames Migratory Fish Report*, Thames Water Authority, London.

THURSTON, G., 1965, *The Great Thames Disaster*, Allen & Unwin, London.

TIMBS, J., 1886, *Clubs and Club Life in London* (new edition), Chatto & Windus, London.

VALENCIENNES, A., 1847, *in* CUVIER, G. and VALENCIENNES, A., *Histoire Naturelle des Poissons*, 20: 340, P. Bertrand, Paris.

VENABLES, G., 1874, *Salmon in the Thames and other Rivers* (3rd edition), Macintosh, London.

WATER POLLUTION RESEARCH LABORATORY, 1964, *Effects of Polluting Discharges on the Tidal Thames*, Water Pollution Research Technical Paper 11, HMSO, London.

WATERS, B., 1964, *Thirteen Rivers to the Thames*, Dent, London.

WHARFE, J. R. and VAN DEN BROEK, W. L. F., 1977, 'Heavy metals in macroinvertebrates and fish from the lower Medway estuary, Kent', *Marine Pollution Bulletin*, 8 (2): 31–4.

WHEELER, A., 1958, 'The fishes of the London area', *London Naturalist* (1957): 80–101.

WHEELER, A., 1960, 'The "common" goby in the London area', *London Naturalist*, 39: 18.

WHEELER, A., 1965, 'Further extensions to the known range of the northern rockling, *Ciliata septentrionalis*', *Journal of the Marine Biological Association of the UK*, 45: 673–8.

WHEELER, A., 1969, *The Fishes of the British Isles and North West Europe*, Macmillan, London.

WHEELER, A., 1969, 'Fish-life and pollution in the Lower Thames: a review and preliminary report', *Biological Conservation*, 2 (1): 25–30.

WHEELER, A., 1977, 'The origin and distribution of the freshwater fishes of the British Isles', *Journal of Biogeography*, 4 (1): 1–24.

WHEELER, A., 1978a, *Key to the Fishes of Northern Europe*, Warne, London.

WHEELER, A., 1978b, 'Hybrids of bleak, *Alburnus alburnus*, and chub, *Leuciscus cephalus* in English rivers', *Journal of Fish Biology*, 13: 467–73.

WHEELER, A., 1978c, 'Rainbow trout in the tidal Thames', *London Naturalist*, 57: 59–60.

WHEELER, A. and MISTAKIDIS, M. N., 1960, 'The skipper (*Scomberesox saurus*) in the southern North Sea and the Thames Estuary', *Nature Lond.*, 188: 334–5.

WILLIAMS, A. C., 1946, *Angling Diversions*, Herbert Jenkins, London.

WILLIAMS, W. P., 1965, 'The population density of four species of freshwater fish, roach (*Rutilus rutilus* (L.)), bleak (*Alburnus alburnus* (L.)), dace (*Leuciscus leuciscus* (L.)) and perch (*Perca fluviatilis* L.) in the River Thames at Reading', *Journal of Animal Ecology*, 34: 173–85.

WILLIAMS, W. P., 1967, 'The growth and mortality of four species of fish in the River Thames at Reading', *Journal of Animal Ecology*, 36: 695–720.

WILSON, D. G., 1977, *The Making of the Middle Thames*, Spurbooks, Bourne End, Bucks.

WOOD, L. B., 1973, 'The condition of London's Rivers in 1971', *Greater London Council, Quarterly Bulletin of the Intelligence Unit*, no. 22: 18–36.

WOOTTON, R. J., 1976, *The Biology of the Sticklebacks*, Academic Press, London.

WRIGHT, D. M., 1971, 'When Londoners fished the Thames', *Country Life*, 11 November 1971: 1309–11.

YARRELL, W., 1828, 'Remarks on some English Fishes, with Notices of three Species, new to the British Fauna', *Zoological Journal*, 4: 137, 465.

YARRELL, W., 1836 (3rd edn 1859), *A History of British Fishes*, 2 vols, John Van Voorst, London.

INDEX

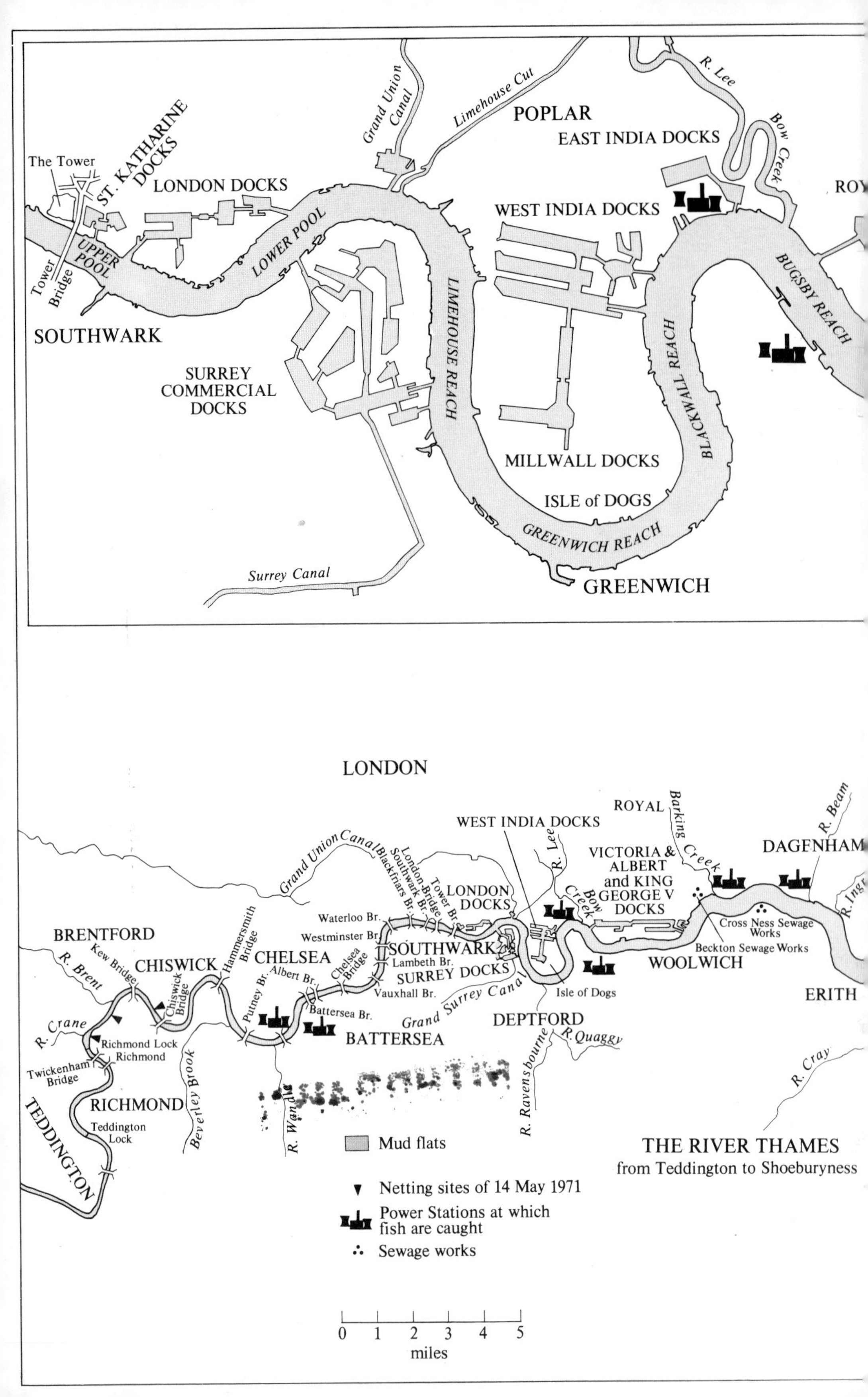

The Tower
ST. KATHARINE DOCKS
LONDON DOCKS
Grand Union Canal
Limehouse Cut
POPLAR
EAST INDIA DOCKS
R. Lee
Bow Creek
WEST INDIA DOCKS
UPPER POOL
LOWER POOL
Tower Bridge
SOUTHWARK
LIMEHOUSE REACH
BLACKWALL REACH
BUGSBY REACH
SURREY COMMERCIAL DOCKS
MILLWALL DOCKS
ISLE of DOGS
GREENWICH REACH
Surrey Canal
GREENWICH
LONDON
ROYAL
WEST INDIA DOCKS
Barking Creek
R. Beam
DAGENHAM
Grand Union Canal
Blackfriars Br.
Southwark Br.
London Bridge
Tower Br.
LONDON DOCKS
R. Lee
VICTORIA & ALBERT and KING GEORGE V DOCKS
Bow Creek
Waterloo Br.
Westminster Br.
SOUTHWARK
Lambeth Br.
SURREY DOCKS
Cross Ness Sewage Works
Beckton Sewage Works
WOOLWICH
BRENTFORD
R. Brent
Kew Bridge
CHISWICK
Hammersmith Bridge
CHELSEA
Albert Br.
Chelsea Bridge
Chiswick Bridge
Putney Br.
Vauxhall Br.
Grand Surrey Canal
Isle of Dogs
ERITH
R. Crane
Battersea Br.
BATTERSEA
DEPTFORD
R. Quaggy
Richmond Lock
Richmond
Twickenham Bridge
R. Cray
RICHMOND
Beverley Brook
R. Wandle
R. Ravensbourne
TEDDINGTON
Teddington Lock
Mud flats
Netting sites of 14 May 1971
Power Stations at which fish are caught
Sewage works
THE RIVER THAMES
from Teddington to Shoeburyness
0 1 2 3 4 5
miles